Visual Participatory Arts Based Research in the City

Visual Participatory Arts Based Research in the City maps ontological, aesthetic and ethic differences between humanist and posthumanist arts-based research while providing insight into methodological orientations to develop arts-based research with frameworks based on process philosophies.

It is the first book on arts-based research which focuses on the city, adopting a posthumanist approach to the assembled nature of urban environments, where agency is distributed across infrastructures, technologies, spaces, things and bodies. Chapters 2 to 8 feature a series of studies, situated in different cities in Europe and the Americas, which outline experiences of movement, inhabitancy, interdependence, collaboration, infrastructuring and sensorial recalibration informed by art practices in film, photography, digital projection, installation, performance and art as social practice. At the core of this book is the idea that aesthetic ecologies of cities do not depend solely on human activity, relying instead on non-logocentric modalities of collective life.

The book is an indispensable tool to researchers, instructors and graduate students in education, the social sciences and the arts aiming to conceive, design and develop projects in arts-based research.

Laura Trafí-Prats is a Senior Lecturer of Childhood Studies and Research Methodologies at Manchester Metropolitan University. Laura's research engages with and responds to children and young people's sensory and material experiences in a variety of contexts including urban spaces, the natural environment and schooling.

Aurelio Castro-Varela is an Assistant Professor of Cultural Pedagogies at the University of Barcelona. His research focuses on learning environments and urban arts-based practices through a new materialist perspective. He is also a member of the research group Esbrina and postdoctoral researcher in the European project Pleasurescapes. Port Cities' Transnational Forces of Integration.

Routledge Studies in Urbanism and the City

For more information about this series, please visit: www.routledge.com/Routledge-Studies-in-Urbanism-and-the-City/book-series/RSUC

Visual Participatory Arts Based Research in the City

Ontology, Aesthetics and Ethics

Edited by Laura Trafí-Prats
and Aurelio Castro-Varela

Routledge
Taylor & Francis Group

LONDON AND NEW YORK

First published 2022
by Routledge
2 Park Square, Milton Park, Abingdon, Oxon OX14 4RN

and by Routledge
605 Third Avenue, New York, NY 10158

Routledge is an imprint of the Taylor & Francis Group, an informa business

© 2022 selection and editorial matter, Laura Trafí-Prats and Aurelio Castro-Varela; individual chapters, the contributors

The right of Laura Trafí-Prats and Aurelio Castro-Varela to be identified as the authors of the editorial material, and of the authors for their individual chapters, has been asserted in accordance with sections 77 and 78 of the Copyright, Designs and Patents Act 1988.

British Library Cataloguing-in-Publication Data
A catalogue record for this book is available from the British Library

Library of Congress Cataloging-in-Publication Data
A catalog record has been requested for this book

ISBN: 978-0-367-46296-3 (hbk)
ISBN: 978-1-032-22109-0 (pbk)
ISBN: 978-1-003-02796-6 (ebk)

DOI: 10.4324/9781003027966

Typeset in Goudy Oldstyle
by SPi Technologies India Pvt Ltd (Straive)

Contents

Figures

Contributors

Portia Cobb is an interdisciplinary artist—a maker of short, single-channel, poetic documentary, auto-ethnographic video and photographic essay. She is also an Associate Professor at the University of Wisconsin-Milwaukee. Her practice includes site-specific collaborative performance as a gesture of expanding the cinematic experience beyond the frame. Her themes are memory and forced forgetting, the politics of place and identity, home and dislocation, and rural and urban spaces.

Brais Estévez is an urban geographer who currently works as an independent researcher. He has done research in Barcelona, where he studied on urban controversies and the politics of public spaces. Between 2017 and the end of 2019 he worked in Salvador, Bahia (Brazil), as a postdoc researcher in the Geography department of Universidade Federal da Bahia. It was in Bahia where his ethnographic work, traditionally focused on grassroots urbanism, met with Black urbanism.

Elizabeth de Freitas is a Professor at Adelphi University. Her research explores innovative data methodologies in the social sciences, anthropological and philosophical investigations of digital life, and cultural-material studies of mathematical practices in complex learning environments. Her work has been funded by the Canada Council for the Arts, the Ontario Arts Council, the US National Science Foundation, and the UK Economic and Social Research Council.

Riikka Hohti (PhD, MA, music) works at the Faculty of Education and the Institute for Sustainable Sciences, University of Helsinki. Riikka has conducted multispecies ethnographic research in educational contexts and written about multispecies childhoods, care relations, digitality, temporality and the affective atmospheres of the Anthropocene. Her current postdoctoral project develops the concept of more-than-human education, with a specific focus on care relations between children and animals.

Rita L. Irwin is a Distinguished University Scholar and Professor of Art Education at The University of British Columbia, Canada. Her research interests have spanned in-service art education, teacher education, socio-cultural issues, curriculum practices across K-12 and informal learning settings, and a/r/tography. She is best known for creating, inquiring and performing a/r/tography in a range of contexts.

Matthew Isherwood is a Doctoral Candidate at The University of British Columbia. His research is located across the fields of aesthetics, queer theory and art education. Matthew's work is focused on exploring how experiences of oppression and prejudice influence the development of one's aesthetic sensibility. In doing so, it looks to the work and lives of queer artists as a way of exploring the idea of *queer energy*, a term used by many queer scholars to describe the disrupting effect of queerness on the status quo.

Rina Kundu Little is an Associate Professor at Texas Tech University and an Education Consultant at the Lynden Sculpture Garden. Her research addresses the politics of knowing and being within educational environments, particularly how ideas and concepts within new materialisms, critical place-conscious education and visual culture studies can be used to critique current practices in museums, research pedagogy, and curriculum and instruction.

Dónal O'Donoghue is Professor of Art and Aesthetic Education at The University of Arkansas and Senior Editor (2019–2021) of *Studies in Art Education*. His research and scholarship focus on gender and contemporary art, specifically contemporary art's educative potential and its capacity to function as a distinct mode of scholarly inquiry and research. In 2018, he was elected as a Distinguished Fellow of the National Art Education Association (NAEA) USA.

David Rousell is Senior Lecturer in Creative Education at RMIT University and a core member of the Creative Agency Lab and Digital Ethnography Research Centre (DERC). He is a Visiting Research Fellow at Manchester Metropolitan University (UK) and an Adjunct Senior Lecturer at Southern Cross University (Australia). David's research combines his scholarship in affect studies, process philosophy and posthumanism with his creative practice as an environmental artist, educator and ethnographer.

Michele Sorensen is an Assistant Professor in the Faculty of Social Work at the University of Regina. Situated in the field of new materialist philosophy, Sorensen studies how non-Indigenous students and social workers might ethically engage the work of decolonization, Indigenization and reconciliation in ways that are respectful of Indigenous worldviews. She is currently researching the potential of aesthetic events for their relational invitation to a broader sociality.

Valerie Triggs is an Associate Professor in Arts Education in the Faculty of Education at the University of Regina. Her research interests include initial art teacher education, teacher education practicum experience and the ecological impacts of art and aesthetic practice. A/r/tography is significant to her for its opportunities in sensitizing selves to curricular potential that is generated in movement and in the repositioning already underway in the infinite encounters between other living and non-living events and entities.

Judit Vidiella is an Associate Professor of Performing Arts Degree at ERAM College (University of Girona, Spain). She is a member of ERAMSCI Research Group and collaborator of ESBRINA Research Group. She has been involved in several research projects related to arts-based research (ABR): Pleasurescapes. Port Cities' Transnational Forces of Integration (2019–2022); Beyond Text (2016–2019); and Performing Practices as a Social Way of Knowledge: A Critical Revision of the Idea of Participation in the Context of Democracies (2014–2017).

Acknowledgements

Visual Participatory Arts Based Research in the City: Ontology, Aesthetics and Ethics has been a labour of love in a time marked by unexpected personal and professional demands. In February 2020, Aurelio Castro-Varela began a visiting fellowship at the Manifold Lab of Biosocial, eco-sensory, and digital studies of learning and behaviour at MMU, where Laura Trafí-Prats works. One of the motivations of Aurelio's visit was to work *tête-a-tête* with Laura in the edition of the book. Such plans got quickly disrupted as Covid-19 swept through Europe in the first months of the year.

By mid-March 2020, UK universities closed buildings and moved their academic activity online. Visiting scholars like Aurelio were asked to return to their countries and long lockdown periods started. Some chapter contributors pulled out to assume new duties and respond to personal emergencies. By April and May, we found ourselves feeling that the book project was moving backward rather than forward, hesitant on how to bring it to fruition. Despite the setback, we managed to attract new chapters at a time when many scholars were going through the countless changes that pandemic time introduced in their professional practice, daily routines and social lives. The successful edition of the book speaks to the contributors' scholarship, commitment and acquiescence. They have written a set of robust and inspiring chapters, which are a testimony of the potential that arts-based research holds for re-assembling collective life in cities. We are humbled and extremely thankful for their talent. In this group of scholars, we want to acknowledge our colleague Rachel Fendler, who started this project with us but could not continue later. Hopefully, there will be new projects in the future where we can converge again.

Faye Leerink, our editor at Routledge, supported the book proposal from the beginning. Her encouragement did not faint when we faced major disruptions. We appreciate the guidance offered by Routledge's editorial team and the flexibility that they have shown.

Laura's career has been touched by the brilliance, friendship and work of scholars, who have inspired her passion for arts-based research. Special kudos go to Fernando Hernández, Charles Garoian, Elizabeth Ellsworth, Kimberly Powell, Dónal O'Donoghue, Maggie MacLure, Elizabeth de Freitas and David Rousell.

Laura's family, Eric and Ingrid need to be hugged a thousand times for always being there through thick and thin, ready to roast some marshmallows. A very special accolade is reserved for her elder mom and dad, Teresa and Pere. Two incredible humans in their late 70s who have been fiercely resilient during inauspicious and lonely times. The four of you deserve the world. Still awaiting there is a toast to raise with Aurelio, who has been a dream partner through this project.

Aurelio has navigated the way to this point thanks to the wisdom, generosity and support of many friends and colleagues in the Cultural Pedagogies Unit at the University of Barcelona, the Esbrina research group, El Solar de la Puri, the Pleasurescapes research project and others. Among them, Fernando Hernández, Brais Estévez, Isaac Marrero, Aida Sánchez de Serdio, Rachel Fendler, Beatriz Revelles, Steve Roberts and Judit Vidiella deserve a special mention. Needless to say, Laura Trafí-Prats also belongs to this list. Working with Laura on this volume has been a real gift, consisting first and foremost in learning from her. There is indeed a toast waiting to celebrate this lucky *sympoiesis*.

Life without Marina would be a mistake, with her it is a joy. Her family (Encarna, Joan, Dàlia and Jude) gladly housed me amid the lockdown and made me feel absolutely at home. And despite the distance, my own family and parents, Esther and Aurelio, are always there for me.

Manchester/Barcelona
June 21, 2021

1 Visual participatory arts-based research in the city

Outlining posthumanist approaches

Laura Trafí-Prats and Aurelio Castro-Varela

1.1 Introduction

The book *Visual Participatory Arts Based Research in the City* engages with practices of inquiry crafted at the intersection of art, theory and the city with the aim to explore and experiment with the relational, sensorial, material potentials of different places and spaces in cities across the world. We propose a journey through nine studies that use concepts and questions emerging from new materialisms, new empiricisms and posthumanist frameworks to consider life in the city as more-than-human and assembled. A central motivation of the book is to widen the imagination around urban life beyond neoliberal and colonial subjectivities and geographies. For this, we suggest a focus on the city as a territory of practice (Amin and Thrift 2017). Deleuze and Guattari (1987) noted that territories are different from a plan or a map because they resist abstraction. Territories are grounded in the earth and express ongoing movement, composition and flow. Territories encompass and provoke felt and embodied relationalities.

Thus, we propose approaching the city as a *posthuman ecology* (Bignall and Braidotti 2019), where things, systems and infrastructures have active powers to augment or decrease the capacities of bodies to move, connect, communicate, think. The city as a posthuman ecology could be thought of as a complex organism that expresses, senses, regulates and controls. In a posthumanist framework, places, things, systems are more than tools and contexts for human action. They are actants that function in relation to other actants creating trajectories and movements in the territory (Bennett 2010). In this posthuman landscape, we see contemporary arts-based research as carrying the potential for inserting virtual openings in the complex architectures, spaces, infrastructures and computational systems that make urban life take on specific forms. Artistic experiments could open the imagination for less passive ways of living with such systems and appropriate them to develop other ways of making *with* them. This can widen the participation in existing urban concerns such as housing, climate change, resource distribution, etc. (Corsín Jiménez 2014). It also can open a space for aesthetic play with things, places and technologies that allow experimenting with times and rhythms (De Freitas et al., 2020) as well as other exercises of re-spatialization that lead to the impossibility of

DOI: 10.4324/9781003027966-1

being pinned down by the spatial, computational, managerial and mediatic designs of the neoliberal city.

In tune with these ideas, our book also recognizes and problematizes the city as a space where there are continued practices of erasure affecting the liveability of subaltern, Black and Brown geographies (McKittrick 2012). The city as a colonial neoliberal space categorizes as *inhabitable* the places of the poor, Black and Brown, so these could be subject to all sorts of interventions exercised from a distance. These are interventions that manipulate but do not feel the city (Simone 2019). To feel the city, one needs to be in the plane of territorial practices, where sensing, mattering and becoming in relation are important aesthetic, ethical and political acts for enacting experiments with modes of collective life that resist such de-humanizing and abstracting gestures. We explore how contemporary art-based research can cultivate such radical forms of inhabitation, movement, composition and re-spatialization.

1.2 A story to take perspective

In this section, we retell the infamous story of artist Mierle Laderman Ukeles' residency at the Department of Sanitation in New York City (DSNY). We do it in the sense that it helps to foreground the idea that collective life in the city could be creatively reinvented through an art practice that opens up and grapples with questions such as, how space is made, maintained and sensed in the city? And how spaces, materialities and bodies entangle in maintenance work? Prior to this residency, Ukeles had become known in the art world for developing the concept of Maintenance Art (1969). Maintenance Art linked practices of care, cleaning, maintenance and feminist politics with conceptual art, performance, documentation, site-specific interventions and collaborations with different institutions in the city, including the DSNY.

Touch Sanitation Performance (1979–1980) was one of Ukeles' initial projects at the DSNY. In the course of 11 months, she shook hands with 8,500 sanitation workers (popularly known as *sanmen*) and thanked them for their work. Maintenance Art turned away from the capitalist notion of *doing something* as connected to ideas of productivity and neoliberal performance and embraced a speculative notion of creativity, which assumed that there is *something doing* in the emergent and undecided socio-material processes of cleaning and managing detritus in the city (Manning et al. 2019). This *something doing* suggests that being in the midst of activities such as conversing, documenting, moving, mapping and performing tasks along sanitation workers can generate modes of sensation, attention, habituation and collective engagement (Manning and Massumi 2014). In turn, this can give form to new sensuous bodies and living styles alternative to the imaginaries and spatial behaviours regulated by the neoliberal and colonialist city.

One of Ukeles' projects that better shows these notions of being amid things consisted in a series of maps based on travelling the city in garbage trucks through designated collection routes and times. In creating these maps, Ukeles learned about sanmen's particular ways of feeling and envisioning the city while handling and moving around types of waste and pollutants encountered in these routes. She became attuned to and highly interested in sanmen's techniques, knowledge and sensorium, and began exporting, experimenting and reconnecting these to new artwork.

Ballet Mécanique for Six Mechanical Sweepers (New York City Art Parade 1983) is possibly one of the most significant examples of how Ukeles adopted and experimented with maintenance techniques. It consists of a choreography that she designed in collaboration with six of the best DSNY sweeper drivers that she met and observed at the DSNY. It mixed concepts of time, movement, ensemble and grace taken from the realm of dance and performance art, with automotive sweeping techniques. The low-rated, invisible even annoying labour of sweeping the streets with a noisy vehicle is rendered as something wondrous, made different, choreographed and artful. This happens through the combination of modes, where choreography as an art concept combines with the dexterous sweeping techniques of the sanitation workers, dancing around iconic avenues and the background of Midtown Manhattan office buildings. In mixing and interpenetrating these different modes, artworks generate creative forces that redistribute the sensible and affective conditions of collective life in the city, showing that the continuous making of subjectivity is interconnected with the continuous making of the world (Ellsworth 2005). Concepts like Maintenance Art and Ukeles' residence at the DSNY also evoke that being in common is not like being the same but "an active belief or ethic that our common being is never given or found but always in the making" (Rajchman 2000, 13). The relational and hybrid nature of the city, any city, provokes that this immanent process of making cannot be reduced to human or social attributes. On the contrary, urban objects, infrastructures and technologies "are the prosthetics that enable subjects to think, act, and feel" (Amin and Thrift 2017, 17; also see, Jacobs 2012).

The DSNY residence and Ukeles' collaborations with sanmen is a good example of how art contributes to intensifying experiences around the city's vital powers. It brings us to consider the city as a medium for art inquiry and speculative thinking (Rousell 2020). In what follows, we try to connect a theory of the city as a complex and agentive assemblage (Amin and Thrift 2017) with understandings of arts-based research as a speculative process of life inquiry (Manning and Massumi 2014).

1.3 For a new materialist posthumanist arts-based research

Our book is not the first source that addresses arts-based research from a new materialist posthumanist approach and with emphasis on place and space. We have been inspired by previous work that has connected with process-based philosophies including Whitehead (1978), Deleuze and Guattari (1987, 1994), Massumi (2002, 2011), Manning (2016, 2020), and Manning and Massumi (2014). This work puts posthuman ecological aesthetics (sensation, feeling, affect, relationality and the minor) at the centre of enquiry. Early on and informed by Massumi's (2002) work on sensation, Ellsworth (2005) formulated a conceptualization of pedagogical inquiry focused on environmental aesthetics. She defined it as "the sensation of coming into relation with the outside world and to the other selves who inhabit and create that world with us" (117). Ellsworth questioned how pedagogical knowledge is tight "to cultural theory's grid of knowledge already known" (120), affirming that such an approach is invalid when thinking in a learning self that is in motion. "The learning self when it is in the making no longer coincides with whatever previously constructed knowledge about the learner we might hold" (121). For the development of her theory, Ellsworth substantially drew on the occurrent arts, architecture and media, all forms of art that develop in the terrain of the city. She sustained that such art forms help to think in terms of bodies that are not fixed but who practice emplacement, inhabitation, construction and thereby make sense in movement and in relation to other bodies. Even more, Ellsworth noted how spatial and architectural arts offer possibilities to explore and experiment with "raw possibilities of movement and sensation that makes possible different corporealities to be expressed" (124).

Along with Ellsworth's (2005) and Kruse and Ellsworth's (2011) work, *Visual Participatory Arts Based Research in the City* connects and extends a body of arts-based research developed through the last decade in the fields of curriculum, pedagogy and social practice that attends to the sensuous, material, more-than-human dimensions of thought. It sees arts-based research as a process of collective experimentation, concerned with how encounters between philosophical concepts, art practices and everyday explorations can generate unanticipated and non-hegemonic relationalities. Additionally, this research has shown an interest in spatial and environmental practices connected to living inquiry (e.g. Irwin and de Cosson 2004), walking (e.g. Springgay 2011; Springgay and Truman 2017, 2019; Lee et al. 2019) and mapping (e.g. Powell 2010; Knight 2016; Rousell 2020) that, while not always engaging directly with the city, are relevant in developing research orientation towards posthuman ecologies.

We are also interested in how this body of research differentiates from the focus of conventional arts-based research on art experiences defined as something exclusively human; humanly created, humanly experienced and about human phenomena (Barone and Eisner 2012). Conventional arts-based research considers that there are aspects of reality that are hidden, and that arts-based research can unearth or make them more accessible (Leavy 2018). However, new materialists and posthumanist arts-based researchers have affirmed that such an approach runs the risk of oversimplifying the empirical potentiality of the arts. Rosiek (2018) sees arts-based research as not only centred on revealing how some discourses frame lifeworlds, but on how the "historical weight and momentum" (636) of existing discourses has material-affective agency. The world is discursive-material and the agency is not only human but things in the world are agents in the creation of realities (Barad 2007; Bennett 2010). Arts-based research can engage with historical weight through art practices that foster "the cultivation of receptivity to a phenomenon or experience, which brings with it a condition of vulnerability to being changed by it" (Rosiek 2018, 640).

For Mazzei (2020), cultivating such receptivity and openness to be changed involves thinking research beyond a logic of deconstruction grounded by humanism and dependent on phenomenological methods in which experience is always of a human subject and refers always to the past. Differing from this, posthumanist arts-based research focuses on a more-than-human notion of experience, not as something existing in the past, but as something that needs to be created by entering in relations with different entities that are more than human. Mazzei (2020) develops such a line of argumentation with a focus on the speculative philosophy of Alfred North Whitehead. Whitehead's notion of experience is not centred on the knowing subject, but in sensing and forming relations. This relational view of sensing resonates with the notion of *receptivity* that Rosiek (2018) discusses in connection to art practices.

In what follows, we offer a series of three possible conceptual clusters that advance more-than-human approaches to arts-based research in the city.

1.4 Conceptual cluster #1: Speculative inquiry

In contrast with conventional arts-based research's reliance on a human subject who accesses profound realities through the aesthetic and sensuous, Whitehead advances a philosophy of the organism driven by an ontological principle according to which subjectivity is not individual but embedded in the entire universe. This means that subjects or entities are not exclusively human, but include any organism, a dog, a rock, a plant, a camera, a street, pollution, an automotive sweeper and so on. The existence of these entities is defined not as discreet but as relational. This means that the subject is not set in advance. For these entities to exist they need to be sensuous (to ingress) and be sensed (prehended) by other entities (Shaviro 2012). An entity *is* because of its capacity to relate to other entities.

Therefore, the focus of speculative inquiry is not on things or beings but on events that bring different entities together in a dynamic process of becoming in which something new is introduced in the world by a nexus of incidents that adhere to one another. Whitehead calls these *occasions*. For Whitehead, each occasion carries a unit of becoming, and when the occasion perishes, this unit of becoming subsists as *datum*, a raw sensuous/affective material that can be taken into subsequent occasions to inform further becomings. For Whitehead, creativity resides in these occasions, but these occasions are punctual moments where entities come together and take on a new form that adds something new to the universe, what Whitehead calls *concrescence*.

This entails two important things. First, that becoming is not ensured, but "punctual and atomistic, and always needs to be repeated or renewed" (Shaviro 2012, 18). Second, objects are not important for what they are, but for being events that accumulate occasions. Events are not things that happen to objects, and objects are not impassive but they are "actively *happening*" (18) even when they appear motionless. Objects like the work *Ballet Mécanique* mentioned earlier are the nexus of actual occasions, a multiplicity of becomings that exist in composition with one another over time, and that likely began through the multiple collaborations and encounters between Ukeles, sanmen, the city, sweepers, detritus and the public. However, as Shaviro (2012, 20) argues, the process of addition of occasions is not continuous: "Nothing comes into being once and for all; and nothing sustains itself in being, as if by inertia or its own inner force". This means that there is not a consequential connection between two occasions like shaking the hands of DSNY workers or mapping garbage trucks routes that lead to the output *Ballet Mécanique*. Existence is relational. Things only come to existence through practices of setting different things in relation, forming different possible joints and connections. Some of these connections harden or acquire concrescence, and we can identify then a new creation. In *Ballet Mécanique*, Ukeles experimentally composed the sweepers with concepts and techniques extracted from choreography and performance, which allowed to experience these vehicles as an urban ballet, thus adding a different visibility in the city landscape and a different valorization of their potential for making space matter.

Thus, and as Stengers (2008) has noted, for Whitehead what means to be and make sense of the world is not a question of subtracting something (processes of capture and interpretation typical of conventional arts-based research) but a matter of adding. This adding involves the construction of ways of paying attention and learning to care for what matters in experience. This is important for arts-based research because it means "that any new creative construction testifies explicitly … to a commitment … that everything we experience must matter" (99). It calls for artists, researchers, and participants to be responsive and response-able to the unique, peculiar, odd aspects that make an experience matter, not to a generalization that "explains away" (99). Therefore, the uniqueness of experience should not be reduced or interpreted with aspects of another experience. It always adds to the world's multiplicity.

In *Process and Reality*, Whitehead (1978) foregrounds the value of concepts for being used in inventive ways to lure new feelings and induce new ways for experience to matter. Concepts are situated at the centre of Whitehead's creationist world and entangled with the experiential nature of thinking. In this way, concepts are redesigned again and again to lure new feelings and to make experience unfold and become eventful in ways that extend the imagination and make certain questions matter (Stengers 2008). Also drawing on Whitehead, Manning and Massumi (2014) have argued that it is through the interpenetration of philosophical concepts and art techniques, what they describe with the term *research-creation*, that art and philosophy can have a generative relation with each other as a way of keeping the creation of concepts connected to what is emergent and to a practice of the event. Thus, art-based research that is engaged in speculative inquiry practice seeks to activate art and philosophy in mutually generative ways to grapple with the invention of new territories of practice and respond to emergent matters of concern in cities, communities, groups.

Donna Haraway (2016), who is a reader of Stengers and Whitehead, offers some examples of how speculative inquiry can engage in remaking collective life in the city. Haraway writes about SF engagements with the matters of the world. SF stands among other things for Speculative Fabulation, a practice of forming creative and experimental collaborations between diverse constituents in our cities: Scientists, artists, citizens, animals, plants, air, technology. Haraway sees art as central in the creation of interdisciplinary and speculative spaces that foster such collaborations in ways that are not competitive or exploitative but that focus on *rendering-capable*.

For Haraway (2016), research is conceived as speculative experiments that foster a multispecies justice of life and death in the context of the many concerns posed by the Anthropocene. (Anthropocene is a term that Haraway herself despises for its emphasis on the Anthropos, proposing others like Capitalocene and Chtulucene). One of the unexpected collaborations that she discusses in depth was devised by digital artists, in collaboration with engineers, pigeon fanciers and pigeons in Southern California. This unusual team worked together sharing knowledge and skills to invent ways of measuring and mapping air pollution in new and helpful ways. The artists and engineers designed a tiny backpack that combined a small GPS, pollution and temperature sensors, cell phone tower capability communication and a SIM card to be carried by the pigeons. Raised and cared by the fanciers, the pigeons were trained in carrying and flying with the backpack. Through this training, the pigeons' flying styles offered feedback to the engineers on the ideal size and weight of the backpack for them to be able to fly at high altitude and acceptable speed so as to monitor air in movement. The cell phone capability allowed to inform the public in real time, streaming about pollution levels and mapping pollution through multiple points. This in turn remade the public's imagination about pollution by showing how it moved across the territory. Haraway has argued how all the players in such a project rendered each other capable through a speculative practice organized as a multispecies collaboration that valued attunement, worlding and response-ability towards experiences of living and dying. The project shows the idea of using the city as a medium for sensing, experimenting and enhancing collective life, and it demonstrates the importance of paying attention to how things come to work together in ways that matter life and experience differently.

1.5 Conceptual cluster #2: Researching with/in force fields and atmospheres

Arts-based research as speculative inquiry engages with what Amin and Thrift (2002, 2017) and Thrift (2008) characterize as the *non-representational aspects* of the city. They refer to urban elements that resist to be explained in terms of human (e.g. intentionality) or social (e.g. language, codes) attributes, and that instead highlight the assembled nature of the city. These are forms of collective urban life that cannot be thought of in terms of parts but only can be engaged as arrangements that envelop many things. It is these more-than-human arrangements that allow for personal and social styles of urban life to take place. Thus, Amin and Thrift (2017, 17) propose to think in terms of *urban agency*, and describe it as a:

> *force field* of relational interactions, [where] hybrid inputs are aligned and made to work through various coupling and amplification devices (e.g. infrastructures, bureaucracies, calculative logics) and a ... general ecology of interactions (e.g. tolerance capacity, population dynamics, flow turbulence).

The notions of force field and urban agency connect with non-representational and non-anthropocentric views of art developed by the philosophy of Deleuze (2003) and Deleuze and Guattari (1987, 1994). Art as a relation of forces separates from art as representation. For example, in *Francis Bacon: The Logic of Sensation*, Deleuze (2003) speaks of painting as a material embodied process "capable of detecting forces that become more and more intense and affects that emit these configurations" (Sauvagnargues 2018, 39). In such a non-representational framework, art is a way of becoming with sensation and of responsibly resonating and mingling with such material forces, allowing the human body to recompose with them or to exist in specific arrangements that can be more capacitating.

A good place where to see how Deleuze and Guattari (1987) bring the question of art as a composition of forces with ecological effects is *A Thousand Plateaus*. There they describe landscapes as made of different invisible and inhospitable forces and write about animals utilizing elements in their bodies and in the terrain to create territories through compositions of plumage, urine, song and movement. A territory is not just individually made but it emerges out of the composition of several milieus that are territorialized, acquiring "a temporal constancy and a spatial range that make it a territorial, or rather territorializing, mark: a signature" (Deleuze and Guattari 1987, 315). Art emerges as the result of entering in relation with ongoing movements of external and material forces and exists in rhythmic variation with them. Art creates a frame that momentarily encircles these forces, making their expressive qualities to "no longer [be] anchored in their 'natural' place but put into the play of sensations" (Grosz 2008, 13). The body becomes a membrane where sensations may be felt and resonate, but sensations are of the world, not of the body. Sensations refer to powers that cannot be lived or perceived directly but that are felt in the body indirectly, in the nervous system, as vibrations or rhythms.

This aesthetic and ecological logic of chaos and variation connects with a thinking of the city as constant flow and heterogeneity. As Boumeester and Radman (2016) note, the city is unpredictable and cannot be captured. In any urban space whatsoever, nothing and nobody ever takes the same path twice, and paths are always open and molecularly affected by different drives and by desires for interactions or stimulations that can make bodies to move in specific directions and also be attracted to change trajectory, detour, slow down and get in random encounters. Paths emerge as a transveral composition of psychical (the body as a sensing membrane), social (the categories and discourses that organize the city as a specific terrain) and environmental dimensions (the earth's material-atmospheric movements). This means that it is impossible to think of the city just with fixed forms or pre-existing molar individualities. The urban experience is always the result of a composition of physical and virtual aspects along with several temporal and spatial vectors, which require a thinking concerned with complex dynamics of mutual constitution and emergence. Considering the relation of the physical with the virtual is important for building non-mimetic cartographies that account for intensity, movement and becoming. "It is not about bringing all sorts of things under a single concept of the city, but about relating each city to the variables that determine its mutation, its becoming" (Boumeester and Radman 2016, 60).

Stewart (2011) has referred to *atmospheric attunements*, to describe how we precisely break through the idea of capturing fixed forms and engage in non-mimetic but sensuous practices of mapping that permit us to think in physical–virtual ecologic conditions, as Boumeester and Radman (2016) propose. Stewart describes atmospheres as made by *force fields*. She writes,

> atmospheric attunements are an intimate, compositional process of dwelling in spaces that bears, gestures, gestates, worlds. Here things matter not because of how they fit in hard politics and social categories but because they have immanent qualities, rhythms, forces, relations and movements.
>
> (445)

Stewart (2011) suggests that perhaps by concentrating on dwelling in, embodying and sensing worlds we can pencil down what is going on. We can notice specificities, minor movements, variables of how one state mutates into another, how becoming takes place in certain urban micro-localities.

Thus, a focus on heterogeneity, force fields and atmospheric attunements renders practices of arts-based research in the city not as an exercise in knowing but in being through experimentation and speculation. In such a framework, art is not a window into worlds that become accessible through thoughtful and sensitive artistic representations, as conventional arts-based research purports. Art is about becoming in the middle of ongoing variation, and where sensation opens spaces to experiment with affects and percepts coming from the outside (sounds, rhythms, planes, volumes, folds, voids) that push sensing beyond human capacities.

In contrast with the world made of birds, ticks, and hermit crabs described in Deleuze and Guattari's (1987) ecological aesthetics, urban sociality is more and more mediated by information systems, data and sensor technology inserted in the built environment which actively shapes ubiquitous sensation (de Freitas et al., 2020). Not being passive to these more-than-human sensuous systems and building capacity to sense atmospheric variation is an important aesthetic, political and methodological gesture for arts-based research to take in the formulation of concerns about spatial and environmental conditions. As McCormack (2018) notes, atmospheres are sensed by and through devices and infrastructures (more on this later) that can create, sense, measure and modify these atmospheres and the environments they shape. In this context, art as an experimentation with sensation can connect with ontological concerns about the agency of things, infrastructures and technologies in shaping both enduring and momentary existential conditions or propagating certain affects.

Artist Tomás Saraceno, who works interdisciplinary across the science of air, art and engineering, shows how art projects can deploy a speculative aesthetic that engages the public in exploring elemental and atmospheric conditions (McCormack 2018). One of his artworks, *Museo Aero Solar*, is an enormous surface of flattened and attached plastic bags fabricated participatively by constituents in the different places where it has been installed. It can take on the air and fly or it can rest on the ground. If the proper atmospheric conditions allow for, it becomes a dome where visitors can move inside and feel the modification of the atmosphere by the colouration of light projected through the bags and the change of temperature. As McCormack (2018) notes, *Museo Aero Solar* is a playful, affective participative experiment that builds imagination on how to distribute capacities of sensing across different bodies and devices and how that can help in stirring public discussion and thinking on the types of infrastructures needed for caring for planetary life. With Saraceno's *Museo Aero Solar* and his entire *Aerocene* project, we see that attention to atmospheres and experimentation with techniques and devices are critical processes to engage collectively in thinking about how atmospheres can dramatically modify but also sustain more diverse forms of life in the city. It enacts the notion of urban life as an unpredictable and located force field of hybrid and coupled interactions between psychic and machinic potentials, and of art as a potential for "creating new worlds and new possibilities of experiencing them" (Grosz 2008, 79). One of the central motifs of Saraceno's work is to democratize the access to experience and experimentation with elemental aesthetics and politics in a context of late capitalism defined by colonial and extractive logics where water, air and soil have been highly manipulated and appropriated. While *Aerocene* delivers a critique of the acute demarcations between the physical body, society, economy and nature (Frichot et al. 2016) performed by the global city as a capitalist hub, its intervention does not stop in the critique, but outlines the possibility of creating a space of transversal connections. This leads us to the third and last conceptual cluster, in which we problematize a body of literature around the right to the city that has linked urban studies with contemporary art. In this, art research is often framed in a political logic that does not always permit to think in the type of complex and more-than-human relationalities described in this second conceptual cluster and through the example of *Aerocene*.

1.6 Conceptual cluster #3: Thinking the right to the city through affect and the aesthetics of infrastructures

The discourse on the right to the city emphasizes the figure of inhabitants in making and remaking the urban in the context of capitalism, further privatization and restricted access to public space. Central to this right is the collective struggle for reimagining and creating another type of city that is not only defined by the primacy of exchange value but for creative activities and non-profit places of encounter (Lefebvre 1996). This means that the right to the city speaks of an *oeuvre* (Lefebvre 1996; also see Duff 2017), enacted through creative process of appropriation and reorientation of urban spaces. Thus, it is the everyday experience of inhabiting a city that offers the potential to anyone or anything to exercise such right, not the fact that one owns, rents or consumes property in the city (Purcell 2014).

The concern on urban experimentation that the right of the city encapsulates has been shared both by urban studies and practices of contemporary art, known as relational aesthetics (Bourriaud 1998), participatory practice (Bishop 2012), art as living form (Thompson 2017) and art as social-engaged practice (Helguera 2011) that have used the city as a laboratory of multiple experimentations, operating at the interstices of art, politics, and everyday life. Projects of socially engaged art have become vehicles for activating communities in relation to public matters of concern (Pasternak 2017). These have included art and architectural projects developed around issues of redistribution of resources and power "between top-down forces of urbanization and bottom up social and ecological networks" (Cruz 2017, 61). As Duff (2017) has noted, speaking of the right to the city in terms of a dualism between macro powers and local resistances has been a common tendency in the literature both of the right to the city and socially engaged art.

While we agree that the right to the city speaks of collective forms of struggle reimagining urban life, we think that methods based on opposition politics and phenomenology are not sufficient to imagine collectivity more radically, broadly and vitaly. Instead, following our posthuman focus, we propose thinking the right to the city by considering how humans are embedded in complex relational systems or infrastructures. In fact, the concept of city as *oeuvre* advanced by Lefebvre (1996) addressed urban transformation as something occurring at the performative level. However, little has been discussed about the specificities of embodied and situated styles of performativity in connection to how the right to the city is materialized (Duff 2017). How is it that bodies and environments acquire new capacities, connections and relationalities that make them become a new collectivity? The fact that the discourse of socially engaged art has focused on hard political forces and dualities (power-resistance) is partly due to a logo-centric view of political participation that does not recognize things and ecologies as a vital aspect in the formation of the demos (Bennett 2010). This view considers aesthetics as political only to the extent that it stays in the realm of the human as well as in connection to artistic practices. Other forms of everyday aesthetics connected to social media (Brunner 2020), vernacular styles (Simone 2019) or infrastructures (Larkin 2013; Berlant 2016) are not seen as key in the creation of the mental, social, affective ecologies of the city.

Current developments in urban studies suggest attention to infrastructures and practices of *infrastructuring* (Corsín Jiménez 2014), affirming that these carry the potential for remaking the city and its forms of sociality at the level of affect (Berlant 2016). While infrastructures are often seen as material forms that facilitate the possibility of an exchange over space and thus thought of as a materialization of the networked character of capitalism and neoliberal governance, current new materialist research insists on approaching them by considering their aesthetic powers (Larkin 2013). Infrastructures create the atmospheric conditions of the city including "sense of temperature, speed, florescence and ideas associated to such conditions" (336–337). Thus, infrastructures work sensationally at the level of the skin and nervous system and before any cognitive register occurs. Attention to infrastructures resituates the politics of the sensible in connection to spatial and temporal sensuous experiences emerging through relations embedded in wider material processes. This is connected to Whitehead's concept of feeling as the capacity of different entities to feel even before knowing or recognizing each other. It is in the possibility that things encounter each other aesthetically that the new forms of collectivity anticipated by the right to the city as *oeuvre* find a starting point.

An artist whose work addresses the potential of infrastructures in generating distributed forms of sensous capacity is Mexican–Canadian Rafael Lozano-Hemmer. His *Border Tuner* (n.d.) enacted a relational architecture that recomposed affectively and infrastructurally an urban territory across the cities of El Paso, Texas and Ciudad Juárez, Chihuahua, situated at each side of the Mexico–United States border. Three interactive stations provided with a speaker, a microphone and voice-activated massive searchlights that were visible from a 10-mile radius, let users engage in dialogue with someone from the stations of the opposite city. To this end, a small dial wheel moved powerful beams of light over the border, which activated communication when colliding with beams coming from the other side. Sound intensity and brightness were also related. The louder the participants spoke, the brighter the light beams would shine. Thus, *Border Tuner* functioned as "a visible 'switchboard'" which brought together a "wide-range of local voices" previously coexisting from the two sides of the "largest bi-national metropolitan area in the western hemisphere" (Lozano-Hemmer n.d.). The installation not only allowed for new connections between El Paso and Ciudad Juárez but also used a range of technological actants and material forms to intensify bodily capacities and make palpable existing connections that the border infrastructure—both an architecture and a legal system enforced by paramilitary and military forces—seeks to continuously contain and outshine.

In recomposing and reassembling technological ecosystems to make them more speculative and sensitively fluctuating (De Freitas et al., 2020), Lozano-Hemmer's works show how infrastructures are both a means for the control of bodies and a means for aesthetic and atmospheric re-assemblings of collective life. His highly networked architectures demand the corporeal involvement of participants in an environmental immediacy with media that fosters shared experience. In the case of *Border Tuner*, beams of light took voices over the border and made people sense the other side as being part of the same territory, *enveloping* both cities under a common atmosphere. As an infrastructure, *Border Tuner* enacted *oeuvres* for feeling the border, making, in turn, its hybridity, fluidity and heterogeneous ecology more palpable.

Focalizing in affects and in the infrastructural materiality of the city as a way of enacting the urban experimentation connected to Lefebvre's (1996) right to the city seems important for the speculative vision of arts-based research that we have argued for in this introduction. It is so because it moves the attention away from the human dualism top-down/bottom-up politics and concentrates on what Bennett (2010) calls a politics centred on the vibrant vitality of things and the personal qualities of affect. Socially engaged art considers that people act because they have become aware of a given issue (Pasternak 2017). Or to say it in a different way, becoming part of an informed collective is what determines their capacity to act. However, in a new materialist perspective focused on affects, a collectivity does not pre-exist the action but is formed performatively by entering in relation with other entities. Experiments on new ways of composing or infrastructuring bodies in the city, like *Aerocene* or *Border Tuner*, are remarkable examples of art-based research as a practice of speculative inquiry in posthuman ecologies.

1.7 Book's organization

The book is organized into three parts each one, respectively, focused on the ontology, aesthetics and ethics of participatory arts-based research in the city. Part 1: Ontological Reorientations further elaborates on one of the central arguments discussed in this introduction, the need of fostering understandings of life in the city beyond anthropocentric frames. So far, we have proposed an idea of the city as a posthuman ecology (Bignall and Braidotti 2019) where collective life involves more-than-human agentive assemblages (Bennett 2010), and where multispecies collaborations among different constituents can take place to generate a multispecies justice of living and dying (Haraway 2016). We have foregrounded the city itself as a relational assemblage where human bodies live in infrastructural and material envelopments that have urban agency and generate the existence of specific styles of urban life (Amin and Thrift 2017). The chapters in Part 1 pick up on some of these ideas and with their own specificities engage in the discussion of more-than-human approaches to urban becomings.

Chapter 1 "Relocating the cinema in the city. The case of El Solar de la Puri", by Aurelio Castro-Varela, is situated in Barcelona, Spain, and more specifically in the neighbourhood of Poble Sec. The chapter proposes a critical intervention in film studies, where the relation of the cinema and the city is mainly theorized in representational terms as the creation of urban views. Castro-Varela suggests that a new materialist ontology reorients such an approach and makes cinema something constitutive in the creation of specific styles of collective existence in the city. Castro-Varela writes,

> that the cinema exists and shelters collective modes of existence within an urban territory is still an unaddressed area of study, and one that requires, too, an account of how the filmic apparatus comes to matter (in the sense of both becoming meaningful and becoming material). This means focusing on the shape that screening technologies and practices take when remaking the city and setting in motion an urban politics of attention, pleasure, and friendship (Amin and Thrift 2002).

With this, the chapter proposes a reframing of Lefebvre's notion of the right to the city as an *oeuvre* through an entity-oriented ontology (McCormack 2018). It focalizes on the study of an open-door cinema as a more-than-human infrastructure with an onto-aesthetic agency that mobilizes different possibilities for feeling and staying at the city as well as a different politics for creating space in the highly gentrified urban geography of Poble Sec.

Chapter 2 "Fred Herzog's affective engagements with things in the city of Vancouver", by Dónal O'Donaghue and Matthew Isherwood, look at the photographic practice of Vancouver-based artist Fred Herzog as an environmental response to the emplaced, temporal and material conditions of the moment. Inspired by the work of queer theorist José Esteban Muñoz (2009), O'Donaghue and Isherwood propose an associative analysis of two of Herzog's photos featuring second-hand stores' windows. This gives rise to a reading of city's spaces that is not limited to their current arrangements, and that opens these spaces "to alternative temporal and spatial maps provided by a perception of past and future affective worlds" (Muñoz 2009, 27). Thus, O'Donaghue and Isherwood illuminate ways to cultivate belonging by evoking a sense of *in-betweenness* in the two images. They argue that *in-betweenness* could be thought of as a "constant movement". Things, like the ones exhibited in the featured second-hand store window, are felt through their vibrant materialities and the evocation of multiple past and future urban trajectories (Bennett 2010).

Part 1 Ontological Reorientations ends with Chapter 3, "Black life and aesthetic sociality in the Subúrbio Ferroviario de Salvador, Bahia", by Brais Estévez. This chapter reflects on the politico-aesthetic activities and urban sociality of Acervo da Laje, a self-managed museum located in a prominent outlying district, called Subúrbio Ferroviario, which city's elites and corporate media have associated with Black death and dispossession. Estévez, a trained geographer with an interest in grassroots processes of urban participation and auto-construction, found himself in Salvador de Bahia in need of a different ontology for recognizing and attuning to the materiality of Black life taking place in and around Acervo da Laje. Thus, the chapter approaches this case study through an onto-aesthetics inspired by the Black radical tradition. More specifically, the chapter focuses on concepts such as *fugitivity* (Moten 2003), the understanding of Blackness as movement, escape and refusal to be reduced to a single thing, and *Black sociality* (Harney and Moten 2013). In dialogue with these concepts, it advances the idea that aesthetics and being are always connected to collective processes of moving and feeling with others that are clandestine, underclass and peripheric. As a result of an ethnographic study, Estévez proposes Acervo da Laje as an arts-based method, the *laje method*, for rematerializing and reimagining Black life in Suburbio Ferroviario through Black study, inhabitability, thingliness and sensuousness.

Part 2 Aesthetic Practices includes two chapters that draw on parallelisms with several issues discussed in this introduction. One of these is that their approach to aesthetics is clearly centred on sensational and affective processes emerging in relation to other bodies, objects and histories in the city. The two chapters highlight how urban objects and spaces are not inert but constitute the nexus of multiple relations and occasions. Additionally, both examine pedagogical projects led by the authors in collaboration with university students, something that emphasizes the role of art and education in creating conditions for other forms of attention, relationality and materiality to take form in the city.

Chapter 4, "Lively pathways: Finding the aesthetic in everyday practice" by Valerie Triggs, Michele Sorensen and Rita L. Irwin discusses an assignment used in a diversity of postgraduate courses. The authors build on their well-known a/r/tographic research practice to formulate thinking about the intersection of aesthetics, pedagogy and research. The assignment promotes an aesthetic involvement of the students with the city through art and writing with the aim of attending to "what they have not previously noticed", and thereby fostering a view of aesthetics as relational, affective and atmospheric (Massumi 2008; Morton 2013, 2013b, 2018). Located in the Canadian cities of Regina and Vancouver, Triggs, Sorensen and Irwin discuss their relation to them through Morton's (2015) concept of *agrilogistics*. Agrilogistics refers to an extractive and mechanistic logic that manipulates environments at a distance and that functions as a matter of fact in the management of the land. Aligning with Morton's critique of this concept, the authors present the assignment as the provision of a space to think outside this logic and to sensitize bodies in the generative aliveness and interdependence of all things in the city. Pedagogically, the chapter connects with Ellsworth's (2005) notion of pedagogical address to make students aware of *something* in their mundane daily routes that "forces one to think" (55).

Chapter 5 "A Hauntological enlivening of the Coma Cros archive through pedagogical inquiry and live performance", by Judit Vidiella, centres on a project of reencountering objects, events and histories at the Coma Cros, an old industrial building in the city of Salt, Spain, that currently houses the undergraduate programme of performing arts where Vidiella teaches. Methodologically, the chapter blends an affective new materialist perspective to archival research (Tamboukou 2014) with performance studies. It delves into the archive of Coma Cros from the time the building was a textile factory, approaching it as an agentic assemblage for experimental and aesthetic encounters with the minor, the silenced and the haunted within archival data (Blackman 2012; Blackman 2019). In collaboration with a group of students, Vidiella utilizes live performances and the theatre of objects, which she both teaches in her university module, as arts-based methods "to resist a literal, linear reading of the archive and to mobilise it as a vibrant aesthetic object and ecology". The encounter with the archive and with former workers from the Coma Cros factory informed the students in the creation and curation of a series of live performances enacted in different spaces of Salt. The performances relived and re-materialized stories and affects in the Coma Cros archive, contaminating the university and the city with them, and thus opening both the building and the surrounding area to non-self-revealing presences and histories.

Part 3 Ethics of Participation examines modes of relation that bring different actors in urban spaces, like the art museum, that are perceived both as hard-to-reach and with the potential to widen representation in such spaces. The two chapters included in this part differ from the assumption accepted in social practice (Bourriaud 1998, Bishop 2012, Thompson 2017, Helguera 2011) that art inquiry starts by activating collectives around already decided notions of what is right and what is wrong. Both chapters suggest that ethical and political concerns take form performatively as bodies are embedded in emplaced, moving, fluid relations in given environments. An art-based logic of speculative inquiry concentrates on the creation of collective events that creatively intensify relationality, affect and sensation among bodies, spaces, technologies and objects (Manning and Massumi 2014). It is when bodies and other entities enter in relation that different concerns about collective life can be formed, experienced and negotiated. Thus, the chapters suggest an ethics of vital materiality, in which we recognize things and bodies being bound by aesthetic-affective styles, public moods and impersonal potentials.

Chapter 7, "The Lynden Sculpture Garden's Call and Response Program: To wonder, encounter, and emplace through the radical Black imagination", by Rina Little and Portia Cobb, discusses the Call and Response Program (CRP) at the Lynden Sculpture Garden in Milwaukee, Wisconsin. The CRP drew its methodology from "a format originating from many African traditions and present in the African diaspora. [It is] often thought of as a pattern where one phrase is heard as commentary in response to another". In Lynden such methodology and museum programme brought Black artists "to co-create in response to artwork made, performances enacted, and materials displayed". Little and Cobb examine how the CRP makes Lynden's land resonate differently with the public. While, it is easy to romanticize the garden as a scape from the gritty and complex racial dynamics in the city, the chapter shows how the CRP cultivated alternative spatial experiences "through material and social practices that are speculatively in tune to a Black presence not recognized before (Nxumalo 2018)". Little and Cobb see this as "an important political gesture in a city like Milwaukee where Blackness and space can be easily connected to dispossession". Theoretically, the chapter is influenced by Black geographies and afrofuturistic aesthetics. Through these lenses, it reflects on how art and social practice can generate "alternative tellings and different spatial imaginaries of the world", in which Black presence can be experienced in more undecided and diverse ways.

Chapter 8, "A poetics of opacity: Towards a new ethics of participation in gallery-based art projects with young people" by Elizabeth de Freitas, Laura Trafí-Prats, David Rousell and Riikka Hohti elaborates on ethics of participation in the context of creative inquiry with young people in art galleries. It centres on a project called *Sensing Time*, which engaged the authors and a group of youth at the Whitworth Art Gallery in Manchester. In this project, the gallery was approached as a creative space for working collaboratively with young people to cultivate new techniques for sensing the complex temporalities of contemporary art, and more specifically William Kentridge's exhibition *Thick Time*. The focus of the chapter is on how the *Sensing Time* project generated a distinctive ethics of participation. For this, the authors draw on the ideas of Édouard Glissant (1928–2011) to reframe participation in terms of a poetics of opacity. They emphasize the ways the project involved "the collective making of a rhizomatic network of errant relays". Thinking with the concept of opacity delivers an alternative way for conceptualizing urban art projects "that might productively stray from standard gallery practices for cultivating youth participation through a politics of identity". *Sensing Time* delivered an ethics of collective becoming based on opacity, through creative practices that grew as "an open proliferation of relays and improvisations". This made intentionality and identification, which are important markers of youth creativity in many museum programmes, to be impossible to trace, favouring instead relationality and hybridization. It delivered a model of participation where the aim was not to cohere with the project but to infuse it with a "fugitive métissage" that opened the network of potential relations.

Chapter 9 is an epilogue that aims to consider the notion of arts-based speculative inquiry in the city through emerging insights and learnings provoked by the experience of Covid-19 and the consequent lockdowns where collectivity in public space was experienced both as life-threatening, and as an opportunity for practices of mutual aid and networks of solidarity.

References

Amin, Ash and Nigel Thrift. 2002. *Cities reimagining the urban*. Cambridge: Blackwell Publishers.

Amin, Ash and Nigel Thrift. 2017. *Seeing like a city*. Cambridge and Malden, MA: Polity Press.

Barad, Karen. 2007. *Meeting the universe halfway: Quantum physics and the entanglement of matter and meaning*. Durham, NC: Duke University Press.

Barone, Tom and Elliot Eisner. 2012. *Arts based research*. Los Angeles: Sage.

Bennett, Jane. 2010. *Vibrant matter: A political ecology of things*. North Carolina: Duke University Press.

Berlant, Lauren. 2016. The commons: Infrastructures for troubling times. *Environment and Planning D: Society and Space* 34, no. 3: 393–419. https://doi.org/10.1177/0263775816645989

Bignall, Simone and Rosi Braidotti. 2019. "Posthuman systems". In *Posthuman ecologies: Complexity and process after deleuze*, edited by Rosi Braidotti and Simone Bignall, 1–16. London: Rowman & Littlefield.

Bishop, Claire. 2012. *Artificial hells: Participatory art and the politics of spectatorship*. London: Verso.

Blackman, Lisa. 2012. *Immaterial bodies. Affect, embodiment, mediation*. London: Sage.

Blackman, Lisa. 2019. *Haunted data. Affect, transmedia, weird science*. London and New York: Bloomsbury.

Boumeester, Marc and Andrej Radman. 2016. "The impredicative city, or what can a Boston square do?". In *Deleuze and the city*, edited by Hélène Frichot, Catherine Gabrielson and Jonathan Metzger, 46–63. Edinburgh: Edinburgh University Press.

Bourriaud, Nicolas. 1998. *Relational aesthetics*. Paris: Les Presses Du Réel.

Brunner, Christoph. 2020 ""Making sense" aesthetic counterpowers in activist media practices". *Conjunctions* 10, no. 7: 3–16.

Corsín Jiménez, Alberto. 2014. "The right to infrastructure: a prototype for open source urbanism". *Environment and Planning D: Society and Space* 32: 342–362. https://doi.org/10.1068/d13077p

Cruz, Teddy. 2017. "Democratizing urbanization and the search for a new civic imagination". In *Living as form socially engaged art from 1991-2011*, edited by Nato Thompsom, 64–71. Cambridge, MA: MIT Press.

De Freitas, Elizabeth, David Rousell, and Nils Jäger. 2020. "Relational architectures and wearable space: Smart schools and the politics of ubiquitous sensation". *Research in Education* 107: 1–23. https://doi.org/10.1177/0034523719883667

Deleuze, Gilles. 2003. *Francis bacon: The logic of sensation*. London: Continuum.

Deleuze, Gilles and Felix Guattari. 1987. *A thousand plateaus. Capitalism and schizophrenia*. Minneapolis: University of Minnesota Press.

Deleuze, Gilles and Felix Guattari. 1994. *What is philosophy?* New York: University of Columbia Press.

Duff, Cameron. 2017. "The affective right to the city". *Transactions of the Institute of British Geographers* 42, no. 4: 516–529.

Ellsworth, Elizabeth. 2005. *Places of learning: Media, architecture, pedagogy*. NY: Routledge.

Frichot, Hélène, Catherine Gabrielson, and Jonathan Metzger. 2016. "Introduction: What a city can do". In *Deleuze and the city*, edited by Hélène Frichot, Catherine Gabrielson and Jonathan Metzger, 1–12. Edinburgh: Edinburgh University Press.

Grosz, Elizabeth. 2008. *Chaos, territory, art: Deleuze and the framing of the earth*. New York: Columbia University Press.

Haraway, Donna. 2016. *Staying with the trouble. Making kin in the Chthulucene*. Durham: Duke University.

Harney, Stefano and Fred Moten. 2013. *The undercommons: Fugitive planning and black study*. New York. Minor Compositions.

Helguera, Pablo. 2011. *Education for socially engaged art: A materials and techniques handbook*. New York: Jorge Pinto Books Inc.

Irwin, Rita and Alex de Cosson (eds.) 2004. *A/r/tography: Rendering self through arts-based living inquiry*. Vancouver: Pacific Educational Press.

Jacobs, Jane. 2012. "Urban geographies 1: Still thinking relationally". *Progress in Human Geography* 36, no. 3: 412–422.

Knight, Linda. 2016. "Playgrounds as sites of radical encounters: Mapping material, affective, spatial, and pedagogical collisions". In *Pedagogical matters: New materialisms and curriculum studies*, edited by Nathan Snaza, Debbie Sonu, Sarah E. Truman and Zofia Zaliwska, 1–12. New York: Peter Lang.

Kruse, Jamie and Elizabeth Ellsworth. 2011. *Geologic city: A field guide to a geoarchitecture of New York*. New York: Smudge Studio.

Larkin, Brian. 2013. "The politics and poetics of infrastructure". *Annual Review of Anthropology* 42: 327–343. https://doi.org/10.1146/annurev-anthro-092412-155522.

Leavy, Patricia. 2018. "Introduction to arts-based research". In *Handbook of arts-based research*, edited by Patricia Leavy, 3–21. New York: The Guilford Press.

Lee, Nicole, Ken Morimoto, Marzieh Mosavarzadeh and Rita Irwin. 2019. Walking propositions: Coming to know a/r/tographically *International Journal of Art & Design Education* 38, no. 3: 681–690. https://doi.org/10.1111/jade.12237

Lefebvre, Henri. 1996. *Writings on cities*. Translated by Eleonore Kofman and Elizabeth Lebas. Oxford: Blackwell.

Lozano-Hemmer, Rafael. n.d. "Border tuner/Sintonizador fronterizo". Accessed May 15, 2021. https://www.lozano-hemmer.com/border_tuner__sintonizador_fronterizo.php

Manning, Erin. 2016. *The minor gesture*. Durham, NC: Duke University Press.

Manning, Erin. 2020. *For a pragmatics of the useless*. Durham, NC: Duke University Press.

Manning, Erin and Brian Massumi. 2014. *Thought in the act: Passages in the ecology of experience*. Minneapolis: University of Minnesota Press.

Manning, Erin, Brian Massumi, and Christoph Brunner. 2019. "Immediation". In *Immediation 1*, edited by Erin Manning, Anna Munster, Bodil Marie Stavning Thomsen, 275–293. London: Open Humanities.

Massumi, Brian. 2002. *Parables of the virtual: Movement, affect, sensation*. Durham, NC: Duke University Press.

Massumi, Brian. 2008. "The thinking-feeling of what happens". *Inflexions* 1, no. 1 (How is Research-Creation? May 2008): 1–40. www.inflexions.org

Massumi, Brian. 2011. *Semblance and event: Activist philosophy and the occurrent arts*. Cambridge, MA: MIT Press.

Mazzei, Lisa. 2020. "Speculative inquiry: Thinking with whitehead". *Qualitative Inquiry* 27, no. 5: 554–566. https://doi.org/10.1177/1077800420934138.

McCormack, Derek P. 2018. *Atmospheric things: On the allure of elemental envelopment*. Durham: Duke University Press.

McKittrick, Katherine. 2012. "On plantations, prisons, and a black sense of place". *Social & Cultural Geography* 12, no. 8: 947–963. https://doi.org/10.1177/1077800420935927.

Morton, Timothy. 2013. *Realist magic: Objects, ontology, causality*. Ann Arbor, MI: University of Michigan Publishing.

Morton, Timothy. 2013b. *Hyperobjects: Philosophy and ecology after the end of the world.* Minneapolis, MN: University of Minnesota Press.

Morton, Timothy. 2015. "What is dark ecology?". *Changing Weathers.* http://www.changingweathers.net/en/episodes/48/what-is-dark-ecology

Morton, Timothy. 2018. *Being ecological.* London, UK: Penguin Random.

Moten, Fred. 2003. *In the break: The aesthetics of the black radical tradition.* Minneapolis: University of Minnesota Press.

Muñoz, José Esteban. 2009. *Cruising utopia: The then and there of queer futurity.* New York: New York University Press.

Nxumalo, Fikile. 2018. "Situating indigenous and black childhoods in the anthropocene." In *Research Handbook on Childhoodnature*, edited by Amy Cutter-Mackenzie, Karen Malone, Elisabeth Barratt Hacking, 2–19. Cham, Switzerland: Springer.

Pasternak, Anne. 2017. "Foreword". In *Living as form socially engaged art from 1991-2011*, edited by Nato Thompsom, 7–15. Cambridge, MA: MIT Press.

Powell, Kimberly. 2010. "Making sense of place. Mapping as a multisensory research method". *Qualitative Inquiry* 16, no. 7: 539–555. https://doi.org/10.1177/1077800410372600

Purcell, Mark. 2014. "Possible worlds: Henri Lefebvre and the right to the city". *Journal of Urban Affairs* 36, no. 1: 141–154. https://doi.org/10.1111/juaf.12034.

Rajchman, John. 2000. "General introduction". In *The pragmatist imagination: Thinking about "things in the making"*, edited by Joan Ockman, 6–15. New York: Princeton Architectural Press.

Rosiek, Jerry. 2018. "Art, agency, and ethics in research: How the new materialisms will require and transform arts-based research". In *Handbook of arts-based research*, edited by Patricia Leavy, 632–648. New York: The Guilford Press.

Rousell, David. 2020. "A map you can walk into: Immersive cartography and the speculative potentials of data". *Qualitative Inquiry* 27, no. 5: 580–597. https://doi.org/10.1177/1077800420935927

Sauvagnargues, Anne. 2018. *Deleuze and art.* London: Bloomsbury.

Shaviro, Steven. 2012. *Without criteria: Kant, whitehead, deleuze, and aesthetics.* Cambridge, MA: MIT Press.

Simone, AbdouMaliq. 2019. *Improvised lives: Rhythms of endurance on an urban south.* Cambridge, UK: Polity Press.

Springgay, Stephanie. 2011. ""The Chinatown Foray" as sensational pedagogy". *Curriculum Inquiry* 41, no. 5: 636–656. https://doi.org/10.1111/j.1467-873X.2011.00565.x

Springgay, Stephanie and Sarah Truman. 2017. "On the need for methods beyond proceduralism: Speculative middles, (in)tensions, and response-ability in research". *Qualitative Inquiry* 24, no. 3: 203–214. https://doi.org/10.1177/1077800417704464

Springgay, Stephanie and Sarah Truman. 2019. "Counterfuturisms and speculative temporalities: Walking research-creation in school". *International Journal of Qualitative Studies in Education* 32, no. 6: 547–559. https://doi.org/10.1080/09518398.2019.1597210

Stengers, Isabel. 2008. "A Constructivist Reading of Process and Reality". *Theory, Culture & Society* 25 (4): 91–110. https://doi.org/10.1177/0263276408091985

Stewart, Kathleen. 2011. "Atmospheric atunements". *Environment and Planning D: Society and Space* 29: 445–453.

Tamboukou, Maria. 2014. "Archival research: unravelling space/time/matter entanglements and fragments". *Qualitative Research* 14, no. 5: 617–633.

Thompson, Nato. (2017). *Living as form socially engaged art from 1991-2011.* Cambridge, MA: MIT Press.

Thrift, Nigel. 2008. *Non-representational theory: Space, politics, affect.* London: Routledge.

Whitehead, Alfred North. 1978. *Process and reality.* New York: The Free Press.

Part I

Ontological Reorientations

2 Relocating the cinema in the city
The case of El Solar de la Puri

Aurelio Castro-Varela

2.1 Introduction

More often than not, cinema has been seen as a continuation of the city by other means. Their historical relationship, frequently remarked as essential to and co-constitutive of both, tends to be grounded on how films are capable of making urban life visible. Cinema is "fantastic at portraying" space as a sphere of mobility, juxtaposition and multiplicity (Lury and Massey 1999, 231), and the modern metropolis is nothing if not a spatial form entailing an intensification of these ontological features. In line with this understanding, a number of scholars have underscored the potential of filmic depictions to project and affect the urban (Barber 2002; Lukinbeal and Zimmermann 2008; Koeck and Roberts 2010), noting how they bring into play a spatial culture illuminating lived metropolitan sites (Shiel and Fitzmaurice 2001) or contribute to forging the symbolic contract which shapes the ideological constructs and material forms of a city (Braester 2010). In one way or another, all these approaches have in common seeing the moving image as "an agent in the building of city views" (Bruno 2008, 15).

This chapter seeks to reconsider the means through which the cinema *goes along with* the city. In what follows, the key point of their association is no longer how films *show* the urban but rather how they are *part* of it. Film studies have conceived too narrowly the concrete materiality that connects both realms, often restricting inquiry into cinematic effects to "considerations of individual subjectivity, consciousness and filmic form" (Pratt and San Juan 2014, 5). Reception and places of exhibition are minor issues in the literature of the field; moreover, the spaces examined are usually too generic (the movie theatre, the art gallery, the internet, etc.) and are considered apart from any socio-atmospheric features or "the film on view" (5). Thus, exploring how the cinema exists and shelters collective modes of existence within an urban territory is still an unaddressed area of study, and one that requires, too, an account of how the filmic apparatus comes to matter (in the sense of both becoming meaningful and becoming material). This means focusing on the shape that screening technologies and practices take when remaking the city and setting in motion an urban politics of attention, pleasure and friendship (Amin and Thrift 2002).

This reconsideration also follows new materialist approaches such as Karen Barad's (2003) onto-epistemic invitation to move from representational concerns

DOI: 10.4324/9781003027966-3

towards performative understandings. Barad sees "representationalism" as the belief in the power of words to mirror "preexisting phenomena", which is also a "metaphysical substrate that supports social constructivist [and] traditional realist [...] beliefs" (802). In contrast, performative alternatives bring to the forefront "important questions of ontology, materiality, and agency" by changing the focus "from questions of correspondence between descriptions and reality [...] to matters of practices/doings/actions" (802). In relation to the city, cinema has also been assigned the role of mirroring its everyday life. Reframing this connection through performativity leads beyond the signification, significance or accuracy of filmic depictions addressing the urban. The urban is in fact the prime location of such depictions. In other words, cinema is not an external cultural force but belongs to, takes place in and configures the city materially. Boundaries between moving images and urban spaces, or screens and streets, can therefore be reimagined or even blurred through other material-discursive enactments. Taking this potentiality into account is already one such enactment.

Later in the chapter, the task of rethinking the cinema as *part* of the city finds a case study in El Solar de la Puri (SP)—a grassroots open-door cinema situated in the neighbourhood of Poble Sec, Barcelona. SP will be understood and discussed as a socio-material entanglement giving rise to unusual expressions of life in and through a very particular location. I suggest that the mode of existence of this filmic infrastructure, which brought together human and nonhuman entities (such as walls, furniture, images, grass, devices, discourses, bodies, shadows, etc.), re-enacted Henri Lefebvre's (1996) notion of the right to the city through the "ontological choreography" (Thompson 2005) of these entities coming together. In so doing, the open-door cinema also worked, as Lefebvre (1996) had claimed for the arts, as a "source and model of *appropriation* of time and space" (173). The screenings taking place at SP—that is, among ruins turned into a site of memory and visual pleasure, as I explain later—led people to experience the urban by means of a (trans)filmic interface. In turn, at the venue the cinema apparatus encountered an atmosphere which embraced communal styles of togetherness.

Infrastructural and aesthetic dimensions overlapped in the SP assemblage. Consequently, my approach to its urban aesthetic politics mobilizes what McCormack (2018) calls an "entity-oriented ontology". This understanding foregrounds "the qualities and properties of nonhuman things or objects" (6) as well as displacing the human from the centre stage of political accounts of reality, in a similar way to the right to the city when infrastructures, rather than people, are theorized *in their own right* as its main source of possibilities (Corsín Jiménez 2014). The SP open-door cinema shared these nonhuman and/or more-than-human ontological bases, and this affected the aesthetic politics at work in its screenings. In this respect, Jacques Rancière's concepts of fiction (Rancière 2009) and the distribution of the sensible (Rancière 2004), both strongly relevant to SP practice, are reconceived through a less anthropocentric lens. Although this practice involved and had effects on human bodies, the means and forces playing leading roles in its mode of existence were mostly socio-material and atmospheric. SP is thus seen here as an "elemental spacetime" that was simultaneously aesthetic, affective and meteorological (McCormack 2018).

The shift from a representational to performative understanding of the ties between the cinema and the city brings to life another displacement in the background of their relationship. This displacement has to do with the centrality of movement and mobility in the representational and with regaining through the performative a sense of location associated to the film apparatus. As I have already remarked, relocating the cinema in the city means first and foremost that the former should not be approached simply as a collection of images about the latter but also as an urban architecture, a form of sociability and a site for learning. Needless to say, this shift involves—as well as expands—that "festival of affects known as a film" (Barthes 1986, 346). In the following section, I discuss some historical precedents and outline the context where this onto-theoretical change of scope, in terms of which SP is explored afterwards, takes place.

2.2 From motion to location

Cinema entails a process of abstraction in which space and time are re-engineered by means of "continuity editing techniques" (Clarke and Doel 2005, 43). This abstraction not only turns the screen into a window on and a frame for other worlds but also converts filmgoers into a hub where these worlds are "made to circulate" (43). From the outset, films offered stationary journeys that took audiences and projected them into the city's movements. From the Lumière brothers' *Arrival of a Train at the Station at La Ciotat* (1895) to the city symphonies—such as *Manhattan* (Paul Strand and Charles Sheeler 1921), *Berlin: The Symphony of a Metropolis* (Walter Ruttmann 1927) or *Man with a Movie Camera* (Dziga Vertov 1929)—the film apparatus provided an inner viewpoint for seeing the daily lives of cities. On the one hand, shooting, lighting and editing, merging into the powers of the *mise-en-scène*, enabled films to grasp urban places and movements. The film shows, on the other hand, took place inside a black box—the movie theatre—which was isolated from the environmental conditions under which those places existed and those movements came into being.

Motion is still considered the umbilical cord between the cinema and the city. It is not only that the camera eye (Vertov 1984) embeds the human view in the rolling flow of the urban, allowing spectators to ride its course of leaps and falls. Besides the film apparatus affording these visual journeys, city movements are also essential in the detail of the images (Clarke and Doel 2005). In this sense, transit and transition are seen as fundamental conditions of both urbanity and screen media (Webber and Wilson 2008). If the city is understandable as a more or less fixed system of spaces and places, including the motions that traverse this structure, the technology of the moving image becomes an "advanced cartographical apparatus" of this "mobility in location" (2). However, the cartographical function of films (see also Hallam and Roberts 2014; Koeck and Roberts 2010, 141–204; Hallam 2010) seems to have obliterated the material potentialities of the cinema apparatus as an urban infrastructure and hence as a mode of location.

Before the invention of cinema, nineteenth-century visual devices—such as panorama or diorama shows and apparatuses—already shared with railroads the ability to dissolve space through time and give access to distant places (Schivelbusch

1986). But the invoked consanguinity of visual and transport technologies (Clarke and Doel 2005, 46), especially that between screenings and trains in moulding optical experiences to the shape of a journey (Kirby 1997), implies much more than similar visualities. It also means that both technologies bring things and bodies together, have an infrastructural basis and are capable of affecting a territory. In this respect, Giuliana Bruno (2008) connects the apparition of films with an emerging network of architectural forms producing a new urban spatio-visuality at the end of the nineteenth century. The moving image arises from a shifting perceptual domain which forges new links between how we move and what we see in the city:

> The luminous aesthetics of panorama paintings and dioramas, the glass architecture of arcades, department stores, pavilions of exhibition halls, glass houses, winter gardens, the electric underground, railways, bridges, powered flight, and skyscrapers incarnated the new geography of modernity. These were all sites of transit. Mobility and light—a form of cinematics— were the essence of these new architectures. By changing the relationship between spatial perception and motion, the new architectures of transit and travel culture prepared the ground for the invention of the moving image, the very epitome of modernity.
>
> (Bruno 2018)

Thus, this new spatio-visuality had manifold locations, even if they moved or were dependent on motions and transits. Bruno (2014) also refers to the screen— both "the fabric and the fabrication of film" (56)—as an *architecture* of light. Hence, her perspective differs from those traditional approaches to urban politics that see the media as a variable which is extrinsic to the metropolis, as if they merely represented a pre-existing world but were not "literally 'built into' urban spaces" (Rodgers et al. 2014, 1059–1060). The existence of films *in location*, as well as the urban assemblages that locate them, come to the forefront in Bruno's spatio-visual model. The city itself turns out to be a "mutable map of modern surfaces" that the urban dweller goes through daily—e.g. "the décor of the hotel lobby", "the shimmering light of the urban arcade" or "the flicking space of the movie theatre" (Bruno 2014, 56). I wonder, however, about the material layers and atmospheric forces involved in a filmic location. Also, are the boundaries between surfaces as clearly defined as Bruno assumes? Beyond this, what I seek to inquire into here is how those boundaries are enacted and how filmic surfaces are energized and spring to life.

Decades before Bruno articulated her theory, Kracauer (1995) approached Berlin's 1920s picture houses as "palaces of distraction" (323). In his view, the style of these "optical fairylands" remodelled the face of the German capital through an "elegant *surface splendour*" (323; author's emphasis). Kracauer relates this refined urban architecture, which avoids any ornamental excess, to the "glittering, revue-like creature" (324) displayed in the auditorium for the show. Added to the smooth architectural envelopment, the "well-wrought grandiosity" of the programme turned these movie houses into modern places of absorption, i.e. places "of transit where a community of strangers gathers to practice the public intimacy of surface encounters" (Bruno 2014, 58). In contrast, in the following sections, I elaborate on SP as a filmic infrastructure which was at odds with experiences of absorption or immersion. The people gathered in and around the open-door cinema did not seek to plunge into an audiovisual spectacle which mobilized them, too, as a mass audience. In fact, SP emerged in spite of, and very often in opposition to, the surrounding urban mobility. In reconfiguring the urban space as a location for coming together, this infrastructure also became detached from other parts of Poble Sec, a neighbourhood unsettled by gentrified forms of leisure. As Kracauer (1998) also remarked, the movie theatre can sometimes serve as a shelter for the "spiritually homeless" (88).

2.3 El Solar de la Puri

Close to Barcelona's historic centre and flanked by the mountain of Montjuïc and Parallel Avenue,[1] the district of Poble Sec has been undergoing gentrification for more than a decade. Local activist groups point to tourism as the main agent of this process, which has caused, among other problems, a rise in rents and prices and the occupation of public space by a plethora of bar terraces, with tourist influxes and mass nightlife activities disrupting the everyday movements of many residents. Local government has also played a part in this transformation, for instance in Puríssima Concepció, an old pedestrian street in the western part of the neighbourhood. A council plan approved in 2000 and finally carried out in 2012 resulted in the demolition of three buildings and the eviction of their inhabitants (see Figure 2.1), most of whom had lived there their whole lives. The plan was to renovate this small plot, building a residence there for international students attending the Theatre Institute, a performing arts college close to that side of the street.

This scheme was never completed due to the financial crisis that broke in 2008. The demolitions left an empty fenced-off plot which remained in these conditions for almost two years (see Figure 2.2). The site became what Gilles Deleuze (1986, 109) calls an "any-space-whatever":

> [This] is a perfectly singular space, which has merely lost its homogeneity, that is, the principle of its metric relations or the connection of its own parts [...]. It is a space of virtual conjunction, grasped as pure locus of the possible.

Although Deleuze uses this concept in *Cinema 1: The Movement-Image* when discussing the affection-images in Robert Bresson's films, I would suggest that it also neatly fits the cinematic transformation that the vacant lot in Puríssima Concepció went through in 2014. A local film collective called Taller de Ficció (TdF) (Fiction Workshop), of which I was a co-founder and member, initiated a participatory process aiming to produce and gather dissenting visual memories of the street. This led to the occupation of the plot in order to create an outdoor cinema. The occupation was rather like making a movie amidst a terrain whose coordinates had yet to be established and which was thus open to being reshaped by any specific affection. The main character of this movie was the cinema itself, which actually *affected* and took the site as its "genetic element" (Deleuze 1986, 110).

Figure 2.1 Puríssima Concepció Street in 2011. Photo by Cristina Silva.

Figure 2.2 The empty fenced-off plot in 2014. Photo by Taller de Ficció.

TdF came together in 2012 around the Poble Sec assembly of the 15M Movement.[2] We took the name from two linked references: Portuguese director Pedro Costa's films about Fontainhas[3] and Jacques Rancière's notion of fiction. Rancière (2004) argues that fiction is neither opposed to reality nor simply consists in "telling stories" (102). Rather, it is a work which seeks to establish "new relations between words and visible forms, speech and writing, a here and an elsewhere, a then and a now" (102). Yet for the TdF group, this fictional work was not restricted to a montage of images and sounds. The whole apparatus through which these images and sounds were projected also came to be a key aspect of the work. We started by creating an audiovisual archive about Puríssima Concepció, bringing together photos, maps, drawings, interviews, documents and our own video recordings. Then we set this archive in motion by sharing it with residents and ex-residents in a two-day workshop. Our idea was to co-compose audiovisual pieces about the street with them. However, most of the participants did not show too much interest in meeting at a social centre to see images and discuss urban issues. After the second session, we realized that another socio-material practice was needed: One that would enable other types of engagement with the street.

After a thorough clean up, our first action at the empty plot—now renamed El Solar de la Puri—was to paint a white screen on the wall and, above it, a sign in red capital letters saying CINEMA (see Figure 2.3). The outdoor screenings started some weeks later, at the end of July 2014. Each session combined dialogue sequences from the archive, basically fragments of interviews, with fictional movies such as *Alice* (Jan Svankmajer 1988), *The Forgotten Ones* (Luis Buñuel 1950) and *My Neighbour Totoro* (Hayao Miyazaki 1988). This blend sought to set in motion a work of fiction

Figure 2.3 The white screen and the red "cinema" sign in the fall of 2014. Photo by Óscar Guillén.

not assigned to any specific genre but consisting in a crossover of sequences in relation to the history of Puríssima. Also, this work did not only involve the screen but
the whole space, turning it into a summer cinema. As one TdF member recalled
during a focus group, Rancière's fiction became a located application:

> The existence of the group was not supported through the discourse as much as
> by ways of doing that gave rise to many events (interviews, screenings, sitting
> in the sun…). It's not a matter of establishing an artificial division between
> spheres, but let's say that the truly important or productive thing was not the
> theoretical definition of fiction, but the practical applications that brought it to
> life. […] The power of a "practical" fiction worked as a uniting force for us and
> between us and other people, such as […] the local residents enjoying "elitist"
> movies or the kids playing and building benches in El Solar de la Puri.

The white screen and the red sign were the only parts of the cinema that stayed
permanently at El Solar. We had to take the other things—projector, computer,
speakers, wires, chairs, etc.—to each session, assembling the filmic apparatus
every time from scratch. Even so, this minor infrastructure worked as the key
to entering the space, redesigning the ruins and transforming them into a community place. A group of kids living in and around the street started visiting
the plot in the evenings to play and make seats with old tyres. Also, in the fall
a group of locals grew a vegetable garden on the right side of the screen. An old
married couple who had lived in one of the demolished buildings joined in with
this activity, probably as a way of making up the loss. Thus, the cinema enabled
other practices to emerge.[4] What was happening here, overall, and how was it
refashioning the urban? In the following section, I theorize this as exercising a
right to the city and, at the same time, expanding the ontology of that right.

2.4 The right to infrastructure

Lefebvre (1996) writes that the right to the city is a "superior form of rights: right
to freedom, to individualization in socialization, to habitat and to inhabit" (173–
174). The right to the city also includes "the right to the oeuvre, to participation
and appropriation (clearly distinct from the right to property)" (174). However,
the industrial metropolis that Lefebvre wrote about in Le Droit à la ville, just
before 1968, is substantially different to present South-European cities such as
Barcelona. Overall, the Poble Sec area is devoted to nightlife and tourism, along
with the service and real-state economies stemming from these. The materialization and performance of SP sought to detach itself from the gentrified spaces of
Poble Sec. It endeavoured to counteract the neoliberal horizon of expectations
by exercising some of the rights mentioned above, especially those concerning
appropriation and the oeuvre. Bearing in mind bearing in mind the differences
between this urban context and previous industrial cities the role that filmic
practices played at SP closely matches Lefebvre's argument on the relationship
between the arts and the right to the city. In his words, the arts should leave
aside "representation, ornamentation and decoration" and embrace "praxis and

poiesis"—i.e. "the art of living in the city as work of art" (Lefebvre 1996, 173).
The work of art, then, consists in offering spatio-temporal forms of producing
urban spaces, as opposed to merely prettifying them.

In this chapter so far, I have been rethinking how cinema comes to matter
in the city. Consequently, I am interested in examining how El Solar de la Puri
inscribed spatio-temporal qualities in the wasteland left after a demolition. However, practising this inscription depends on an ontology which goes unnoticed in
Lefebvre's original understanding. The notion of the right to the city has been
taken up again and renewed in the context of the austerity policies deriving
from the 2008 financial crisis. Researchers have posed questions such as how this
right is linked to do-it-yourself urban design (Douglas 2013, 2015), what kind of
right it is supposed to be (Attoh 2011) or how it should not be understood as an
"individual liberty to access urban resources" but as a way of "changing ourselves
by changing the city" (Harvey 2008, 23; c.f. Purcell 2014). In addition to these
approaches, Alberto Corsín Jiménez (2014) has rethought the right to the city as
a socio-material practice oriented towards changing the overall ecology of everyday urban life. An open-source urban project in another occupied plot, *El Campo
de la Cebada* in Madrid, led him to outline a right "to infrastructure" which is at
odds with human-centred entitlements:

> It is neither a *right to* infrastructure, nor an infrastructure *made right*. Rather, the
> right to infrastructure allows us to escape the human–nonhuman and episte-
> mology–ontology dichotomies altogether by opening up the agential work of
> infrastructures as a source (an open source) of possibilities *in their own right*. [...].
> [This right] is 'collected' somewhat differently, as it gathers materials, devices,
> appliances, media systems, interfaces, and social relations in a dance of graph-
> ematic concatenations. It is a right incarnated in and deployed through very
> specific (open source) sociotechnical designs, interventions, and affordances.
>
> (Corsín Jiménez 2014, 343–358)

Lefebvre's wish for an "ephemeral city" (1996)—i.e. a city that responds to "the
perpetual *oeuvre* of the inhabitants, themselves mobile and mobilized for and by
this *oeuvre*" (173)—is turned by Corsín Jiménez's approach (Corsín Jiménez 2014)
into a more technical question: "What would a city look like if its infrastructures
were designed, built, certified, and managed by its residents?" (342). Also, in the
phrase "the right to infrastructure" the last word—infrastructure—reads as a verb
and not as a noun (357). In tune with this re-conceptualization, I suggest that
the SP's white screen and red sign *re-infrastructured* the street, giving rise to a new
mode of togetherness. However, the open-air cinema did not seek to impose a
comprehensive plan on the space or a sense of clarity about what could be done
there. The right enacted through the filmic infrastructure was rather a "right to
indifference [...] as the capacity to 'let things be', without having to incessantly
make a decision about what is the 'right' way or not" (Simone 2016, 200).

The film screenings gathered things, people and images on the basis of an *indifferent* atmosphere. This is to say that their connection did not respond to a dialogic
approach, as TdF had tried at first in line with the 15M Movement methodology,

but to the very process of coming together for a while, for several reasons. If devices, wires, seats and other elements were reassembled in order to make up a film apparatus for a couple of hours, the audience was also heterogeneous and motivated by diverse, complementary desires. Some people came to see the movie or perhaps a fragment of the archive because they or their relatives appeared in it. For others, the cinema meant hanging out, having a drink or some food, and enjoying a cool place in summertime. And the children saw it as an opportunity for playing in a green, open space before the movie started. In fact, a group of kids was constantly present but only sat through the projection once. This was during a session which combined the screening of *My Neighbour Totoro* with an interview with two sisters, aged 10 and 17, raised in Puríssima Concepció and still living there with their family.

This indifferent atmosphere brought about diverse attachments to the filmic infrastructure, even when these attachments were not exclusively filmic. Also, SP aesthetic practice did not only rely on the images projected on the screen. According to Larkin (2013), infrastructures have an aesthetic agency of their own. They are "matter that enables the movement of other matter" so that their "peculiar ontology" lies in the fact of being "things and also the relation between things" (329). This ontological feature also implies that infrastructures shape sensitive variables of urban networks and hence, the ambient environment of a city's everyday life: "Softness, hardness, the noise of a city, its brightness, the feeling of being hot or cold are all sensorial experiences regulated by infrastructures" (337). This argument is close to the Aristotelian concept of *aesthesis* in the sense that this mode of aesthetics does not consist in the appreciation of artworks but in a "bodily reaction to lived reality" (336).

This open-door cinema slotted into a demolition site that was also a green space felt cool, friendly, affordable and promising for a summer night. It was indeed a location for disconnecting collectively from the increasingly gentrified atmosphere of Poble Sec. Thus, the open-air screen not only brought to life the right to the city by *infrastructuring* an empty plot. It also propagated a right to indifference by letting things go. This second feature worked environmentally rather than simply infrastructurally, which leads me ultimately to conceive of the SP cinema as an urban transfilmic envelopment.

2.5   A transfilmic envelopment

During a projection at El Solar de la Puri, the off-screen was not completely dark but often disturbed by the lights of a car parking, kids going back and forth or the arrival of spectators. The screen—contained between a façade and a supporting wall, both still standing, like a papier-mâché stage (see Figure 2.4)—flickered at nightfall with shadows and lights from the street. In short, the cinema enveloped the plot as much as it was enveloped by the atmosphere of Puríssima Concepció.

McCormack (2018) describes an envelopment as "the condition of being immersed within an atmosphere" (4). It "can be sensed" but not always, or not entirely. Moreover, it is also a process, "a kind of 'extrusive' shaping of things in relation to an atmospheric milieu" (5). To see it only as the former, McCormack adds, "is to risk making too clear-cut and static a distinction between entities and atmospheres" (5). In SP

Figure 2.4 The poster that announced the first film session at El Solar del Puri on 5 July 2014. Design by Cristina Silva.

the entanglement of entities and the atmospheric milieu, and of screen and street, constituted the core of the cinema and the sense of its location. The atmospheric milieu of the venue was that of the city but reshaped by a screening practice.

The envelopment provided by the film apparatus has often been compared to a black box. Roland Barthes (1986) puts it in this way when explaining that in the darkness of the movie theatre ("anonymous, populated, numerous") "lies the very fascination of the film (any film)" (346). The dark is the "substance of reverie" and "the 'color' of a diffused eroticism", both of which impel bodies to slide down into the seats "as if into a bed" (346). This is also the allure or enchantment of an architecture that limits sources of light—its main elemental force—to the "dancing beam of the projector" (347). Outdoor cinemas look like an opaque cube only under the cover of night. Although SP apparently reproduced this model, its mode of attunement was not based on isolation from, but adjacency to, a previously existing urban environment. Rather than a cube, the site worked as a lightbox.

In McCormack's terms (2018), the space-time of palpability provided by the SP open-air cinema was not just focused on the films on view, but also on the screen itself as a green shoot on a recovered plot of land. In other words, Puríssima Concepció Street and the traces of its erased past were not avoided but instead made palpable through the filmic infrastructure. An envelopment emerges from the encounter between atmospheric forces and the allure of entities, entailing "distinctive powers and capacities to act, to sense and to move" (McCormack 2018, 34). The envelopment of El Solar de la Puri merits the term *trans-filmic* since the urban was remade by the screen, and vice versa, in tune with the prefix *trans-*, which means both through and beyond. Going to the movies at SP meant taking part in an urban occupation, inhabiting a common piece of land, staying *al fresco* for an evening, being close to unknown but friendly people, making time simply to enjoy images, sensing the neighbourhood from a slight remove and many other *doings*.

Additionally, I would suggest that this affective engagement echoes but is not quite the same as Rancière's concept of the redistribution of the sensible (Rancière 2004). In his words, the redistribution of the sensible involves a politics of artistic practices consisting in reconfiguring "the landscape of the visible [...] [and] the relationship between doing, making, being, seeing, and saying" (45). But the ontological character of SP's sensible landscape was not just visual or textual. Moreover, it was not exclusively subjective, but above all atmospheric. It is true that its transfilmic envelopment could be felt "in bodies of different kinds" (McCormack 2018, 4). However, atmospheres are not only important in relation to the "affective and sensory capacities of humans" (11) but also have a meteorological basis, in the sense of depending on elemental patterns and variations. The SP cinema had this too, since it needed, for example, summer nights to come alive. Poble Sec residents wanted to go there because it was cool and hospitable, calm and totally affordable, dreamy and out of the way. In this respect, SP reshaped urban elemental forces through and in addition to the screenings.

SP reconfigured the landscape of the neighbourhood, which is sensible as much as infrastructural, not only through films affecting people but on its own or, in other words, as a location. The aesthetic politics at work there may have contributed to the "formation of political subjects that challenge the given distribution of the sensible" (Rancière 2004, 40), but the matter of this politics was nonhuman and atmospheric. Political subjectivization is theorized by Rancière (2004) as a "disincorporation", which means among other things a bodily displacement undoing a pre-existing "distribution of roles, territories and languages" (40). This displacement modifies the sensory perception of what is common and how spaces and occupations are linked in a community. This is also the role of art when constructing "fictions [or] *material* rearrangements of signs and images" (39). Literary locutions, for instance nineteenth-century novels, "widen gaps" and divert people from their *natural* purpose "by the powers of words" (40). However, what if aesthetic powers are not simply those of words or images but of atmospheric things? Can these powers go beyond the centrality of "man" as the only political and literary "animal" (40)? The location of SP was in itself a disincorporation resulting from bringing together "questions of affect, emotion, feeling, and mood with a concern for the [...] elemental milieu in which entities are enveloped" (McCormack 2018, 20).

This is not to say that the films shown at El Solar did not matter. Of course, they did. Each screening addressed a specific issue about Puríssima Concepció by interweaving audiovisual fragments from the TdF archive (interviews, photos, documents, etc.) with fictional movies. In the inaugural session, we showed a short TdF film about the occupation of the plot, which had been an intensively manual and at times even archaeological process. This was followed by Jan Svankmajer's *Alice*, in which stop-motion techniques emphasized the handmade ethos and opened a door to the imaginary. In the second session, we began with a TdF interview with an ex-resident explaining the pedagogical relationship he had established with some young squatters in his building two years before the demolitions. This was paired with Luis Buñuel's *The Forgotten Ones*, a surrealistic movie about marginal, uncared-for Mexican youngsters. In the third session, we presented a TdF video featuring two young sisters and residents of Puríssima Concepció imagining the things that El Solar could ideally have, swimming pool included. This video was followed by Miyazaki's *My Neighbour Totoro*, which blurs the boundaries between reality and fantasy. Although these filmic encounters mattered very much, so did their emplacement in a transfilmic envelopment in which the historicity and environment of the street could be sensed. In so doing, the cinema also became a location for being together in urban homelessness.

2.6 Conclusions: A relocation

In this chapter, I have theorized the case of El Solar de la Puri not as a template to be applied but as an onto-aesthetic invitation to reconsider how the cinema comes to matter in the city. If urban movement and the depiction of a *mobility in location* have been the most recognizable nexus between the moving image and the modern metropolis, I suggest that their historical relationship also offers other, more performative potentialities. Performativity is in fact a contestation of the excessive power granted to language and/or images in determining what is real (Barad 2003). Not just a builder of city views, the cinema is also an architecture of light, a form of spatio-visuality usually linked to other forms of transit. How does it work, however, in and as a location? The leftovers of the demolished houses in Puríssima Concepció Street furnished the surface where a screening infrastructure, as a result of a practical work of fiction, came into being; and this becoming also calls into question aesthetic approaches which, as in the case of Rancière's (2004, 2009), still focus on signs and consider human subjectivities before and apart from more-than-human forces.

Ruins, gaps and any-space(s)-whatever rather than elegant, well-defined surfaces, may provide a relocation for present-day screens. A contemporary right to the city finds in urban scraps suitable scenarios for designing, constructing and managing common infrastructures. A simple layer of paint gave rise to the SP outdoor cinema, which re-enacted that right. But this envelopment had no other plan than to let things go around it. It was not just about films but also about a mode of coming together. It became transfilmic by displaying an indifferent atmosphere, a togetherness which combined diverging forms of being together (Figures 2.5 and 2.6).

Figure 2.5 El Solar de la Puri in the evening before the start of a filmic session, in May 2015. Photo by Taller de Ficció.

Figure 2.6 The transfilmic envelopment of SP during a screening in 2017. Photo by Laboratorio Reversible.

Notes

1 Montjuïc is a broad shallow hill overlooking Poble Sec and Barcelona harbour. It has a high concentration of tourists due to its various museums, the '92 Olympic stadium and commanding views over the city. On the other side of Poble Sec, Parallel Avenue draws a 2-kilometre line from the docks to Espanya Square and forms a border with the Raval and Sant Antoni neighbourhoods. Parallel Avenue was well known for its proliferation of music halls and other entertainment venues in the first half of the twentieth century.

2 In June 2011, the *Indignados* or the 15M Movement in Barcelona brought the occupation of Plaça Catalunya (the city's central square) to a close and dispersed to the neighbourhoods, creating local area assemblies. One of these was formed in Poble Sec. A range of working committees—on the economy, communications, public space, exchange of goods, etc.—was set up to organize protests and spread information to the everyday life of the neighbourhood.

3 Fontainhas is the name of a Lisbon neighbourhood that no longer exists. Demolished at the end of the twentieth century, it was a squalid outlying area, a mix of casbah and shantytown, where a population of poor Portuguese people and Cape Verdean immigrants lived hand-to-mouth in the '70s. From *In Vanda's Room* (2000) to *Vitalina Varela* (2019), Pedro Costa's films have addressed both Fontainhas' disappearance and its haunting legacy in collaboration with ex-residents. In Costa's words, a "desire for fiction" motivated this encounter from the beginning: "I went there [...] to record the loveliest movie ever about the room [of Vanda], the neighbourhood, Portugal, the world—that's not a documentary ambition" (Neyrat 2008, 47).

4 I use a past tense referring to El Solar de la Puri because the fieldwork of my research was carried out between the summer and the fall of 2014. However, the site is still in operation and continues to put on screenings, concerts, lunch meetings and festivals for local residents, amongst other activities. The vegetable garden also remains on the right side of the cinema.

References

Amin, Ash and Nigel Thrift. 2002. *Cities. Reimagining the urban.* Cambridge: Blackwell Publishers.

Attoh, Kafui. 2011. "What kind of right is the right to the city?" *Progress in Human Geography* 35: 669–685. https://doi.org/10.1177/0309132510394706

Barad, Karen. 2003. "Posthumanist performativity: Toward an understanding of how matter comes to matter". *Signs: Journal of Women in Culture and Society* 28, no. 3 (Spring): 801–831. https://doi.org/10.1086/345321

Barber, Stephen. 2002. *Projected cities: Cinema and urban space.* London: Reaktion Books.

Barthes, Roland. 1986. *The rustle of language.* Translated by Richard Howard. New York: Hill and Wang.

Braester, Yomi. 2010. *Painting the city red: Chinese cinema and the urban context.* Durham: Duke University Press.

Bruno, Giuliana. 2008. "Motion and emotion: Film and the urban fabric". In *Cities in transition. The moving image and the modern metropolis*, edited by Andrew Webber and Emma Wilson, 14–28. London: Wallflowers press.

Bruno, Giuliana. 2014. *The book surface: Matters of aesthetics, materiality, and media.* Chicago: University of Chicago Press.

Bruno, Giuliana. 2018. "Architecture and the moving image: A haptic journey from pre- to post-cinema". *La Furia Humana* 34. http://www.lafuriaumana.it/index.php/67-archive/lfu-34/781-giuliana-bruno-architecture-and-the-moving-image-a-haptic-journey-from-pre-to-post-cinema

Clarke, David B. and Marcus A. Doel. 2005. "Engineering space and time: Moving pictures and motionless trips". *Journal of Historical Geography* 31: 41–60. https://doi.org/10.1016/j.jhg.2003.08.022

Corsín Jiménez, Alberto. 2014. "The right to infrastructure: A prototype for open source urbanism". *Environment and Planning D: Society and Space* 32: 342–362. https://doi.org/10.1068/d13077p

Deleuze, Gilles. 1986. *Cinema 1. The movement-image*. Translated by Hugh Tomlinson and Barbara Habberjam. Minneapolis: University of Minnesota Press.

Douglas, Gordon C.C. 2013. "Do-it-yourself urban design: The social practice of informal "improvement" through unauthorized alteration". *City & Community* 13, no. 1: 5–25. https://doi.org/10.1111/cico.12029

Douglas, Gordon C.C. 2015. "The formalities of informal improvement: Technical and scholarly knowledge at work in do-it-yourself urban design". *Journal of Urbanism: International Research on Placemaking and Urban Sustainability* 9, no. 2: 117–134. https://dx.doi.org/10.1080/17549175.2015.1029508

Hallam, Julia. 2010. "Film, space and place: Researching a city in film". *New Review of Film and Television Studies* 8, no. 3: 277–296. https://doi.org/10.1080/17400309.2010.499768

Hallam, Julia and Les Roberts. 2014. "Mapping the city in film". In *Toward spatial humanities: Historical gis and spatial history*, edited by Ian N. Gregory and Alistair Geddes, 143–171. Indiana: Indiana University Press.

Harvey, David. 2008. "The right to the city". *New Left Review* 53: 23–40.

Kirby, Lynn. 1997. *Parallel tracks: The railroad and silent cinema*. Exeter: University of Exeter Press.

Koeck, Richard and Les Roberts, eds. 2010. *The city and the moving image. Urban projections*. London: Palgrave MacMillan.

Kracauer, Sigfried. 1995. *The mass ornament: Weimer essays*. Translated by Thomas Y. Levin. Cambridge, MA: Harvard University Press.

Kracauer, Sigfried. 1998. *The salaried masses. Duty and distraction in Weimar Germany*. Translated by Quintin Hoare. London: Verso.

Larkin, Brian. 2013. "The politics and poetics of infrastructure". *Annual Review of Anthropology* 42: 327–343. https://doi.org/10.1146/annurev-anthro-092412-155522

Lefebvre, Henri. 1996. *Writings on cities*. Translated by Eleonore Kofman and Elizabeth Lebas. Oxford: Blackwell.

Lukinbeal, Chris and Stefan Zimmermann, eds. 2008. *The geography of cinema. A cinematic world*. Stuttgart: Franz Steiner Verlag.

Lury, Karen and Doreen Massey. 1999. "Making connections". *Screen* 40, no. 3 (autumn): 229–238. https://doi.org/10.1093/screen/40.3.229

McCormack, Derek P. 2018. *Atmospheric things: On the allure of elemental envelopment*. Durham: Duke University Press.

Neyrat, Cyril. 2008. "Un mirlo dorado, un ramo de flores y una cuchara de plata. Conversación con Pedro Costa". In *Pedro Costa* [DVD]. Barcelona: Intermedio.

Pratt, Geraldine and Rose Marie San Juan. 2014. *Film and urban space. Critical possibilities*. Edinburgh: Edinburgh University Press.

Purcell, Mark. 2014. "Possible worlds: Henri Lefebvre and the right to the city". *Journal of Urban Affairs* 36, no. 1: 141–154. https://doi.org/10.1111/juaf.12034

Rancière, Jacques. 2004. *The politics of aesthetics. The distribution of the sensible*. Translated by Gabriel Rockhill. London, New York: Continuum.

Rancière, Jacques. 2009. *The Emancipated Spectator*. Translated by Gregory Elliott. London: Verso.

Rodgers, Scott and Clive Barnet, Allan Cochrane. 2014. "Media practices and urban politics: Conceptualizing the powers of the media-urban nexus". *Environment and Planning D: Society and Space* 32: 1054–1070. https://doi.org/10.1068/d13157p

Schivelbusch, Wolfgang. 1986. *The railway journey: The industrialization of time and space in the nineteenth century*. Oxford: Oxford University Press.

Shiel, Mark and Tony Fitzmaurice, eds. 2001. *Cinema and the city: Film and urban societies in a global context*. Oxford: Blackwell.

Simone, AbdouMaliq. 2016. "Urbanity and Generic Blackness". *Theory, Culture and Society* 37, no. 7–8: 183–203. https://doi.org/10.1177/0263276416636203

Thompson, Charis. 2005. *Making parents: The ontological choreography of reproductive technologies*. Cambridge: MIT Press.

Vertov, Dziga. 1984. *Kino-eye. The writings of Dziga Vertov*. Translated by Kevin O'Brien. Berkeley, Los Angeles, London: University of California Press.

Webber, Andrew and Emma Wilson, eds. 2008. *Cities in transition. The moving image and the modern metropolis*. London: Wallflowers press.

3 Fred Herzog's affective engagements with things in the city of Vancouver

Dónal O'Donoghue and Matthew Isherwood

3.1 Introduction

In this chapter, we turn to two artworks made in Vancouver by the Vancouver-based artist Fred Herzog to consider the ways in which these artworks animate thinking about practices of exchange in the city. Operating from the belief that there is a second-hand quality to all of our encounters in and with the city, given that our encounters are always framed and shaped by things known and unknown, we study Herzog's *Second Hand Store Boy* (1959) and *Second Hand Shop, Cordova Street* (1961) using an associative mode of analysis, informed by the work of José Esteban Muñoz (2009). We do this to understand what these artworks reveal about urban economic practices that tend to be easily dismissed, overlooked or viewed as inconsequential. We suggest that an associative analysis and reading of Herzog's work—a process that we explain in greater detail below—point to "the city" as a complicated composition expressed through a series of interconnected elements, each integral to the shape and texture that renders the city recognizable.

Herzog's work seems an appropriate choice for such a project. From his arrival in Vancouver in the 1950s[1] to his death in 2019, he spent many decades studying the city by photographing it—thus producing an extensive visual archive of the city of Vancouver. His visual account of the city—its neighbourhoods, occupants, forces and flows of desire, attraction, attachment, alienation and capital—presents it as a place that shows up differently at different times of the day, in the sun, rain, snow and wind; a place that is always becoming different, while in part remaining the same, as it transitions out of what it has been to become something else. Revealing how people dwell in the city and documenting the visual signs of urban development and economic change, Herzog's images suggest that he was interested in the social and affective environment of the city as much as its built quality. Works, such as *Handshake* (1960) and *Magazine Man* (1959), are a testament to his attention to the city as a place in which intimate exchanges can occur. But his images of Vancouver were also shaped by what was available to picture at the time in which these images were produced. For example, the existence, appearance and practices of working-class neighbourhoods in Vancouver together with the activities that occurred in the entertainment district seemed to be of particular interest to him. Many of his photographs are of such

DOI: 10.4324/9781003027966-4

places. Studying the images he made in these city quarters reveals his curiosity about (attraction perhaps) the city's "grittiness"—a quality he found in the city's "streets and storefronts, hangouts and hideouts, plazas and parlours, backyards and work yards, voyeurs and loiterers" (Herzog, cited in Cheung 2019, np).[2]

Equally, the opportunities that Herzog had to make these images (mainly at weekends when free from work commitments) and the materials that he had access to and chose to use (Kodachrome film with its distinct capacity to produce vivid colour) shape what eventually got pictured. The medium of photography itself enabled Herzog to get close to qualities of the city—qualities that he desired, and the aspects of the city that he remained curious about, wished to be part of or avoid. The act of making a photograph permits one to look at things from which one might ordinarily avert one's gaze (O'Donoghue 2019). Coming to live in Vancouver from elsewhere perhaps influenced in part what he noticed, what he failed to notice, what he pursued, what he ignored and what he brought to our attention through his work. Milroy and Herzog (2011) would refer to this as an "émigré sensibility"—understood as a sensibility enabled and cultivated by his outsider status and his curiosity about the things that were unfamiliar to him (Milroy and Herzog, 2011).

We acknowledge these aspects of Herzog's life and inclinations at the outset of our discussion, not with the intention to amplify the authorial authority of the artist in understanding his images but to acknowledge that these images were made within specific conditions, times and places. In recognizing this, we also acknowledge that an artwork is made not once but many times over by all those who show an interest in it (Bourdieu 1996) and therefore it is never reducible to the artist's intention. As Michael Craig-Martin says, "Works of art are not straightforward embodiments of their initiating ideas" (Craig-Martin 2015, 259). Herzog's acts of framing the phenomena that caught his attention, composing them into an image, making quick decisions concerning what ought to be included and what ought to be excluded resemble what Kathleen Stewart describes as "a mode of production... that throws itself together in a moment [that] is already *there* as a *potential*—a *something* waiting to happen in disparate and incommensurate objects, registers, circulations, and publics" (Stewart 2008, 72). Thus, Herzog's response to the city, his entanglement in it, his framing and decision-making processes, all fix an image of the city and present it to viewers as an account of how the city appeared at a given moment. In these framing and decision-making acts, Herzog collected aspects of the city together that are inventions of sorts, given that photographs are records of what is there to be recorded, but in the act of framing of photographs, in the act of making it, the photographer makes decisions that greatly shape what will ultimately be delivered as a photograph. For Herzog, as he explained,

> the word 'city' for me became more than the intersection of many roads. As early as 1956, I saw the city's many manifestations in icons, archetypes and bipolar contrasts. I was both actor and flâneur because I wanted to know what the city feels like.
>
> (Herzog, interview with Arnold 2006)

In short, then, while most of Herzog's images of Vancouver are photographs that depict city streets, back alleys, back yards, vacant lots, overpopulated sidewalks, neon signs, billboards, cafes, doorways, street corners, barbershops and window displays, they are vivid accounts of the varied and complex things, moments, punctums, affects and publics that existed or were waiting to come into existence in the city of Vancouver at that time. Hence, our interest in them. Further, while his photographs are historical images in some respects—as they document the city that once was and is no longer the same, at least in terms of appearance— they animate curiosities about how the city was organized in the past and continues to be organized in the present. They reveal ways in which the city supported certain forms of exchange and participation in collective life. They picture the indifference that some of Vancouver's inhabitants displayed towards others, an indifference that was adopted as a mode of survival—a "blasé attitude" (Simmel as cited Sennett 2018, 54).

3.2 Associative analysis and reading

As mentioned in our opening paragraph, for our work in this chapter, we borrow the concept of associative analysis and reading from the work of queer theorist José Esteban Muñoz. In *Cruising Utopia*, Muñoz describes associative analysis as a "mode of analysis that leaps between one historical site and the present" (Muñoz 2009, 3). To that end, Muñoz explains:

> my writing brings in my own personal experience as another way to ground historical queer sites with lived queer experience. My intention in this aspect of the writing is not simply to wax anecdotally but, instead, to reach for other modes of associative argumentation and evidencing (3).

Muñoz goes on to provide an example of how this form of analysis is pursued by him. He writes,

> when considering the work of a contemporary club performer such as Kevin Aviance, I engage a poem by Elizabeth Bishop and a personal recollection about movement and gender identity. When looking at Kevin McCarty's photographs of contemporary queer and punk bars, I consider accounts about pre-Stonewall gay bars in Ohio and my personal story about growing up queer and punk in suburban Miami (4).

We turn to Muñoz's methodology of associative analysis seduced by its potential to expand and complicate that which appears given and known in a particular situation, event, time or place. His methodology complicates what tends to pass as the ordinary, the natural, the way things are and the ways things ought to be. His approach does so in an effort to imagine such situations, events or phenomena otherwise. To that end, his mode of associative analysis contributes to the production of a different set of formations without devoting unnecessary energy to dismantling existing ones. His approach is not too dissimilar from Stewart's way of building understanding developed in her books, *A Space on the Side of the Road* (1996) and *Ordinary Affects* (2007), especially her commitment to opening gaps in "the seemingly 'straight' story of a place in an effort to take notice of the 'spaces' within it" (Stewart 1996, 7). Herzog's photographs seem to do the same. Muñoz's methodology is also informed by his notion that "the present is not enough … for queers and other people who do not feel the privilege of majoritarian belonging" (Muñoz 2009, 27). He argues that "the present must be known in relation to the alternative temporal and spatial maps provided by a perception of past and future affective worlds" (27). Thus, his approach seems particularly productive in our project insofar as the act of reading the present through past and future places, one tends to find oneself in a state of being in-between. This quality of in-betweenness might have analytical relevance and scope in seeking to understand the importance of the objects of the city that we consider in this chapter. Elspeth Probyn describes this "in-betweenness" as a state of belonging "not in some deep authentic way but belonging in constant movement" (Probyn 1996, 19). In other words, objects in the city both animate and are animated by affects of belonging.

Imagine, for example, how one can feel connected to certain neighbourhoods in a city, to particular buildings and spaces. Now imagine the inevitable change that all such spaces go through and the feeling of displacement that can follow. Imagine straying too far into the "wrong side" of town, of suddenly feeling an uneasy sense of being unwelcome or under threat. These signs point to the changing atmospherics of the material city. Something like this can be read throughout Vancouver photographer and artist Jeff Wall's writing on Herzog's work, where he affectionately describes the aesthetic qualities of "old Vancouver" as marked by a "gracious air of appropriateness" against the "disappointing city we live in today" (Wall 2017, 34). While still acknowledging Wall's perspective, some are drawn to the sleekness of Vancouver's newer aesthetic in ways others are not (Coupland, 2000). What does this mean? Perhaps, first, That the materiality of the city is always developing and that while some find belonging in its everchanging appearance and quality, others may find themselves alienated or displaced entirely. Second, that Herzog's photographs are "lucid document(s) of change" (Company 2017, 6) that affect one's sense of belonging and more. Considered this way, the images of second-hand stores become ways of exploring this phenomenon.

An associative reading of Herzog's images of second-hand stores helps one arrive at a sense of "the city" as a composition expressed through a series of interconnected elements in motion, each integral to the shape and texture that renders parts of the city recognizable, accounting not only for its past but future also. By photographing these stores and the objects therein, Herzog also recorded a moment of becoming for the city that may not have been noticed at the time. It could also be suggested that Herzog's photographs relay an economy of belonging within the city. For example, these second-hand stores were primarily located in working-class neighbourhoods. Thus, they are documents of the affective structure of the city, how it organizes people from different classes, races, ethnicities and reflects such organization. Other images from Herzog reveal how information becomes framed and conveyed through billboards, shop signs, neon signs, newspaper stands and shop windows. By attuning himself to the forms and rhythms of the city's materiality, Herzog captures the force of objects in his images, affirming how they relate to issues of belonging, commerce, change and survival. It is these qualities that we wish to explore further in images of second-hand stores.

3.3 An associative reading of Herzog's *Second Hand Store Boy* and *Second Hand Shop, Cordova Street*

Second-hand stores were one of the first things that Herzog photographed shortly after arriving in Vancouver (Herzog et al. 2011). The concept of selling used goods in a store devoted to that purpose was unfamiliar to Herzog. In his native Germany, he told Robert Enright (2011), such practices did not exist. For Herzog, it would seem that second-hand stores in Vancouver were places that revealed much about life in the city. They offered an insight into what people owned, used, sold off and wished to acquire. They functioned as places where people could go to sell items to make ends meet. They revealed what people were willing to buy and part with. They revealed, in part, economies of the city (especially in working-class areas). They revealed the politics of some city dwellers (Sennett 2018).

Like the city, in a second-hand store of Herzog's Vancouver, many different objects appeared together, oftentimes arranged in accordance with function, actual or perceived, arranged in accordance with an image of a consumer yet to appear. Like the second-hand store, the city, too, has a coherence and logic in its structure and organization. It tends to be organized around the streams of function and usability (Sennett 2018). Shopping districts tend to be separate from financial districts. Residential areas seem to have a greater number of certain types of stores—grocery stores, laundromats, recreation centres and other types of daily convenience. Perhaps, like the second-hand store, the city in its physical form and idea embodies a commitment to the life yet to be lived. In the city, we encounter things that have already been made, are already in motion, having come from somewhere and going elsewhere. Like objects in the second-hand store, the city reveals layers of lives lived within it, in pursuit of its freedoms, cognizant of its limitations and seduced by the possibilities of belonging that it hints at. As suggested in the introduction to this chapter, to some extent, a second-hand quality to all of our encounters in and with the city, given that our encounters are always framed and shaped by things known and unknown in advance of those meetings with things, desires and seductions.

Herzog's *Second Hand Store Boy* (1959) is a photograph of the window display of one of the many second-hand stores in Vancouver in the late 1950s.[3] Standing in front of the window display, and thus included within the frame of the photograph, but positioned to its right, is a young freckle-faced boy, no more than 10 years of age, it seems. With his hands crossed behind his back, he stares directly at the camera, at Herzog, and at the viewer yet to come. The young boy almost blends in with the background display that he stands before. The plaid patterned shirt that he is wearing matches the pattern of objects on the window behind him. Worn over a spotless white T-shirt, his plaid patterned shirt is tucked inside his oversized trousers held up by a brown belt that seems older than the boy himself. Behind him and behind the glass window of the store, a collection of objects sit on or dangle from shelves. At first glance, these objects do not seem particularly valuable or rare, or of vintage quality. They seem to be simply used objects that are offered for sale. Objects that others might be interested in acquiring. Objects that some others no longer need and were willing to part with. Watches. Many of them. All hanging from a copper bar that runs the entire length of the window. Pocket watches and wristwatches. They all seem to be men's watches. Also, on display in the window are small silver boxes, snuff boxes, perhaps. Two travel clocks, each with their own case. Scissors dangle over the upper shelf. Binoculars. Several binoculars. They are heaped one on top of the other. Cameras, tripods, camera cases, large finger rings some still in their presentation boxes. Knives. Many of them. Most are in their leather cases. Cases that were made to protect them and to protect us from them. Silver medals hang from ribbons. Like the ones you would see on a decorated soldier. A small bourbon-coloured leather hand satchel appears on the window also. One wonders how many of these objects were given or received as an acknowledgement of love, care, achievement or accomplishment. They carry with them histories, stories and forms of attachments impossible to tell by merely looking at them.

In taking such a close-up image of the second-hand shop, especially in *Second Hand Store Boy*, or perhaps by being called by the display itself to give it central prominence in the image being made, Herzog eliminates everything else that surrounds the window of the shop and presents it and the young boy who stands in front of it facing the camera as the primary subject of his photograph. And, he juxtaposes the young boy alongside items that seem on visual inspection older than the boy himself. Like the objects on display, the young boy awaits a life yet to come. Like the objects he stands before, one cannot help not imagine the future that might lie ahead of the boy. Like the objects on display, those he stands before, it seems that there is a life script already in place for this young boy. Whether he will follow it or not, one can never know. While he seems to blend into the pattern of objects in the background, he, too, stands out. He is young. And being young, he has a life before him. Many of the objects on the window have been used. They are in search of another life. Not yet ready to be thrown out.

Following the work of Jane Bennett (2010), one could suggest that the objects on display in *Second Hand Store Boy*, behind the window, not only disclose something about how life was lived and organized in the city of Vancouver at that time—the existence of second-hand stores revealed something about the economic hardships of the city—and thus function as historical documents of sorts, but they also present as animate things. Not passive things. As animate things, they animate other things. Appearing as things already alive and waiting to be put to use in ways that reflect, exceed, and yet never fully corresponding to the intentions of their makers, users and the uses to which they have been previously put. Suspended from shelves that run horizontally across the shop window, these objects tend to occupy a suspended state of sorts or a state of transition. Having lived part of their lives with others from whom they are now separated, waiting to be sold to another and to be put to some future use, they are removed from active life. By being taken out of circulation, they await a future yet to come. But, in waiting, these things are neither dead nor inert matter. As brute matter, they have been categorized as this or that kind of object, given a name, a use, a value, even, and acquired a history through use and attachment. And yet, as matter, these things on display are teeming with life, with many possibilities for animating thought about how others live in the city, how others are forced or enticed to sell these objects or buy them because they cannot afford or wish to buy new ones. They suggest perhaps a longing for how things make life different, better even. These things on display, which attract attention, arouse curiosity, fuel desire, present as objects with a life not yet fully lived. Due to their material quality, some will continue to function for many years to come, and some will fade in colour as a result of exposure to the sunlight. Some will cast shadows on others when the sun shines from one direction and not another. The light that will fall on these objects will bring some of them to prominence while casting some others in shadow. Some objects on that window will reflect light, while others will absorb it. They are put on display to serve a purpose that they exceed—to attract a potential buyer, to arouse curiosity as they seem to have done for the young boy who stands before them or for Herzog, who has photographed them.

Thinking with Bennett (2010), these objects might be understood as lively things—things that have been in the hands of city dwellers, things that have worked the hands of those dwellers and things that have changed hands. As lively things, they, too, have changed through the uses to which they have been put. But, even more likely, they have also changed the things that they have encountered. Together with those who have used them, and the conditions of their use, these things, now on display on the window of a second-hand shop, have brought about, enabled or frustrated change of some kind—be it a change in the material qualities of other materialities or their own materiality. For instance, we could imagine the knives that we see on display as having been used to cut or slice, divide or slash other materials and materialities. They may have sliced bread or cut a loaf in two. Thus, we might say that the knife is a piece of dangerous matter. Depending on its sharpness, and the force with which it is applied to the surfaces it encounters, it can cut through many different materialities. It can inflict wounds. It can threaten life. Used with enough force and repeatedly, it could take the life of its user, of another human, nonhuman animal or being. Most likely, the portable clocks have accompanied people on their travels and have awoken them from their sleep. Thus, while they tell the time, they also do something else. Thus, thinking with Bennett (2010), it could be said that these objects have enabled and participated in the will of humans, or they have impeded or frustrated that will. They have extended to some degree their user's capacity to act, to alter things, to add something to the world and to make something in the world. Now it seems they are at rest, waiting on the window to be noticed again by future users. Even in their waiting, they animate Herzog. They capture his attention. They prompt him to act. He makes a photograph of them.

In this respect, then, the second-hand shop of Herzog's Vancouver, much like the city of Vancouver at the time, is a place full of things that are lively and alive, things that others can imagine uses for, and things that are encountered alongside other things that one might not expect to encounter, intentionally seek out or anticipate their effects. Things that have been used once before. Things that have histories and stories attached to them. Things that are waiting to be used again. Things that attract attention, hold it, and generate desire in the one who looks, in the one who takes notice, in the one who imagines a future for the object within one's gaze. Things with force. Things that cannot be noticed. Things that call the attention of passers-by. Things that demand attention, imagination and companionship. Things that have shaped the life and emergence of the city and are now being shaped by the city in return. It is a place that perhaps prompts consideration of how objects enable one to extend one's capacity to act. Working with them and in correspondence with their capacities, we achieve things. We are able to accomplish things that we might not otherwise be amplished.

Second Hand Shop, Cordova St. (1961) is a photograph shot from outside of the store.[4] Picturing objects commonly associated with the lumberjack trade such as axes, a pickaxe, a sledgehammer and several spades. All are chained together like dangerous animals. In this image, Herzog draws our attention to a label pinned to the handle of the sledgehammer that declares a "sacrifice price" on these items. Images of Roman temples come to mind, wishful offerings to the gods. In the window, a poster declares a "special" on new logger boots, "small sizes" only. Above, running the length of the window hangs an assortment of belts. Many are thick, robust and lacking any fine details, suited to the doing of manual labour. Others stand apart: Ornately patterned leather belts adorned with gold and silver buckles, jewels and other fashionable decoration. Nestled amongst this assortment of things are objects of recreation. A single tennis racket, a golf club, a chessboard. Hidden further back amongst the coats and jackets, one can make out the body of a guitar, while directly below a violin is perched upon a music stand. These objects point us to the mosaic-like diversity of the city, where items of high culture, leisure and labour come together, suggesting the texture of Vancouver's cultural milieu. The objects that appear within the frame—objects of the city—seem to demand a response—a set of questions that extend beyond the desire to imagine their users and wearers. Were these objects a burden? A necessity? An extended part of those who used them? Furthermore, as they wait to be taken up anew, to inhabit a life beyond the life they once lived, they also ask one where they may yet go?

In *Second Hand Shop, Cordova St.*, the shop window, full of the tools of a livelihood under threat, evokes "a state of being in between where things are neither fully present nor absent but linger and echo in a simultaneous lack/excess" (Stewart 1996, 67). The tools, for example, functioned in particular ways, partly pointing, without doing so, to the people that held and used them in their daily lives. One may find it difficult to think of the lumberjack without the tools to fell a tree or the calluses that those tools left on their hands through time and use. Yet, having been sold, these tools now await being used in other ways, held in place by Herzog's camera like a question mark. Why are they being sold? Who previously owned them, and why had they parted with them? Was this simply how these tools could continue to contribute to the paying of rent or the purchase of food? In the image, is there a sense of a way of life reaching its end in the city?

Picking up on the theme of being in-between, the image could be said to represent a moment arrested in the ongoing narrative of Vancouver's Gastown area, now home to some of the city's most vulnerable people. Initially, the Lower Eastside was a central hub of city activity, but as businesses moved westward, the area changed dramatically. More recently, many of the buildings have begun a process of gentrification, with the presence of restaurants, shops and apartments emerging back into the area. The Woodwards building, for example, used to be an iconic commercial building for that area before it closed in the early 1980s. Today, it is an ensemble of modern apartments, social housing, a centre for the arts and retail outlets. Yet, despite token gestures of sensitivity towards the area's history, born mainly from protests and resistance from locals, the Woodwards and similar developments have displaced a large number of people from the area. The second-hand shop in the image was located just a few blocks away from these sites of change. It is another example of how, through associative reading, the work of Herzog becomes a way to explore "the active role of *nonhuman* materials in public life" (Bennett 2010, 2).

Second Hand Shop, Cordova St. appears to show the moment when objects, human and nonhuman, come together to form and inform the life direction of one another. The way in which nonhuman objects: Shops, personal belongings, homes, and places of community pull together in powerful groupings, informing and forming the lives of those connected to them. In turn, these places also find themselves changed. The "grittiness" of these areas, which drew Herzog's attention so vividly, often garners a very different reaction from many Vancouver residents, affirming that one's connection to such spaces is felt in "their encompassing atmosphere—their mood—to which we connect" (Bennett 2012, 61). Recent gentrification and increasing geographic density have pushed many of the human inhabitants from the areas Herzog photographed and onto the streets or surrounding areas. This, in turn, has raised the visibility of homelessness and drug addiction, causing stress and community tension (McElroy 2019). Spaces like Vancouver's Oppenheimer Park have become repurposed, offering temporary belonging to those displaced. Make-shift markets, where personal belongings and discarded items are sold and traded between homeless residents regularly occur, a sub-economy that continues and enables lives in new and different ways. This gives some indication to how materials exude a "not-quite-human force that addle(s) and alter(s) human and other bodies" (Bennett 2010, 2).

In viewing Herzog's images of second-hand stores, it is likely that one considers the practice of selling and buying second-hand items, which, as Brenda Parker and Rachel Weber (2013, 1096) say, has long been a tradition in urban areas. Second-hand shops have long "been a feature of urban retail landscapes", they write (1096). Because they have already been used, objects sold as second-hand are typically offered at a much lower price than they would cost to acquire as first-hand objects. Others have already made use of them and that shapes their value when they re-enter the market. The market for second-hand goods extends the life cycle of consumer objects in general and make them available to others, especially to those others who are unlikely to be able to afford to buy them new. Markets for second-hand goods reduce the overall consumption of new goods together with the "associated negative employment and environmental effects associated with this consumption" (1097). And second-hand shops, Parker and Weber say, "often contribute to vibrant and diverse neighborhood economies and milieu" (1097).

As suggested by Herzog's photographs, the second-hand shop offers many different objects for sale that do not share much in common apart from the fact that they have been previously used and owned. Offering a collection of objects, which have already passed through the hands of others, for sale, the second-hand shop of the past and the thrift stores, markets and consignment shops of present offer a physical space for people to spend time in the city—to spend time browsing, looking through, or sorting things that others have parted with; things that have been removed from circulation, things waiting to be put back into circulation under a different set of conditions. Once purchased and reused, the object goes back into circulation, perhaps returning to its original or intended use. In other situations, an entirely new set of uses may be imagined for the object. In some cases, such as the logging tools, perhaps their original purpose no longer exists as a viable means of life for the object, an occurrence that, as with many electronic devices, is becoming a more and more frequent fate for modern objects. For some, the very act of searching for and finding second-hand goods may bring a certain excitement, animating them through the hope of finding something of great value for a fraction of its original cost.

While a predominantly urban practice, for the most part, the decision to buy second-hand goods is now a deliberate choice, not one made out of necessity as it tended to be in the past. One typically chooses to buy pre-owned goods as an outward sign of their commitment to sustainability, environmental consciousness and eco-friendliness or as an effort to promote the value of reusing and recycling things that have already been produced (Sorensen and Jorgensen 2019). But, as noted, this has not always been the case. In the past, the act of buying second-hand goods carried with it a certain amount of stigma. It was thought that only those who could not afford to buy new goods bought them second-hand, while those in need of money sold objects that they had previously acquired and used. Regardless of the conditions that lead one to buy second-hand items, the second-hand shop permits one to acquire things that they might not ordinarily acquire. Thus, the second-hand shop becomes this place of possibility, as much a reminder for some of what one cannot have if it was not offered in its used state. In their study of luxury consumption in the second-hand market, Turunen and Leipämaa-Leskinen (2015) make a distinction between those who go shopping for and buy second-hand goods and those who seek out and acquire vintage goods. The latter group is motivated by nostalgia and the desire to acquire unique goods. The former group, on the other hand, seems motivated by frugality, sustainability and eco-friendliness.

Thus, Herzog's images of second-hand shops do more than merely represent them visually and capture what they looked like to the eye. They have the potential to raise the type of awareness, consciousness and ways of thinking about how we encounter the things that others put in place for us. Thus, these objects imagine us to some degree.

3.4 Concluding thoughts

Considering the objects in Herzog's second-hand store windows help us to also consider how the city, too, is made from objects that have been built with specific intentions in mind—objects forgotten, lost and rediscovered, only to become taken up differently or repurposed in other ways. How areas of a city can be considered as assemblages of things and people that are at once the producers and products of the systems they entail. Considered this way, the "city" becomes an unstable construct understood more as a moment that was always already there as a potential "something" waiting to happen (Stewart 2008, 72). Not some stolid entity, it pulses with the animate energy of the things that compose it. What connects this idea of the "city" to the images taken by Herzog, considered here through those of second-hand stores, is not how they help one understand a definitive idea of Vancouver, but in how they capture a moment of assemblage in all its vital animacy. In other words, one comes to see the city through Herzog's eyes. In turn, one remembers his eyes were directed by the objects that drew his attention. These objects, suspended in the image, become animate once again on viewing, capable of taking on new meaning through personal context and association. Thus, one begins a poesis "of a something snapping into place" (83). This act, as Stewart tells us, is something

> that literally can't be seen as a simple repository of systemic effects imposed on an innocent world but has to be traced through the generative modalities of impulses, daydreams, ways of relating, distractions, strategies, failures, encounters, and worldings of all kinds (72).

The effect of such a practice of reading is that one arrives at a sense of being in a state of constant production. Thus, rather than evidence of the artists' intention or documentation of a "past" Vancouver, the images tell of a constant production and potentiality. One that does not begin or end with the taking of the photograph or the death of the artist. At the age of 88, Herzog passed in September of 2019. One might believe that, with his death, his images and what they potentially mean also ended. There will, for example, never be a new Herzog image taken of Vancouver, and that is sobering in its finality. From here, the work could easily become viewed as the record of a man's life work and artistic intention. However, this chapter suggests this is not and never should be the case. Much like the city he photographed, Herzog's images live on, awaiting to be taken up in new and vital ways, much like the objects they captured.

In an interview with Sarah Milroy, Herzog was quoted as saying that "a secondhand store window, that is a psychogram right there. It's a very concreate showcase: a list of who we were and how we made our living" (2011, 13). The promise of the object, how one sees their relationship with it, changes what it may tell us about the city of which it is a part. This kind of associative analysis that we conducted here considers the objects of the city as animate things that reflect and impact the way the city develops, and thus the way we live our lives. How areas can generate a sense of belonging for some while excluding others.

How objects, having been released from a former purpose, go on to live in an animate space in which they find the means to negotiate a future of possibility. Here, in this in-between space, these objects relate tales of their "efficacy" within the city, indicating their creative agency and "a capacity to make something new appear or occur" (Bennett 2010, 31). Read this way, Herzog's photographs become like the objects collected in the second-hand stores. They are a way of exploring the city's animate materiality and animate material themselves, caught between Vancouver's past and a future that is yet to make itself known.

Notes

1 Herzog was born in Stuttgart, Germany and immigrated to Canada in 1952, eventually settling in Vancouver.
2 For Herzog, "the thing that street photographers hope to discover [is] … the disorderly vitality of the street; the street people on the corners and plazas, in billiard parlours, pubs and stores, where shoppers, voyeurs and loiterers feel at home" (Herzog 2005, 160–161).
3 See the photo here: https://www.equinoxgallery.com/our-artists/fred-herzog/#work_gallery_bd9a62e76cad7e742ae8eb70b2235948
4 See the photo here: https://www.equinoxgallery.com/our-artists/fred-herzog/#work_gallery_c6f6f35659dde2f73748c297766a4d41

References

Arnold, Grant. 2006. "Interview with Fred Herzog." *Douglas and McIntyre*. Last modified August 15, 2006. douglas-mcintyre.com/interview/19.

Bennett, Jane. 2010. *Vibrant matter: A political ecology of things*. North Carolina: Duke University Press.

Bennett, Jill. 2012. *Practical aesthetics: Events, affects and art after 9/11*. New York: Palgrave Macmillan.

Bourdieu, Pierre. 1996. *The rules of art: Genesis and structure of the literary field*. Stanford, CA: Stanford University Press.

Cheung, Chris. 2019. "The life of Fred Herzog: Vancouver's beloved photographer." *The Tyee*. Last modified September 13, 2019. thetyee.ca/Culture/2019/09/13/Fred-Herzog-Vancouver-Beloved-Photographer-Life-Death.

Coupland, Doulas. 2000. *City of glass*. British Columbia: Douglas & McIntyre.

Craig-Martin, M. (2015). *On being an artist*. Art/Books

Enright, Robert. 2011. Colour his world: The photography of Fred Herzog. *Border Crossings* 30, no. 3.

Herzog, Fred. 2005. "Exploring Vancouver in the Fifties and Sixties." *West Coast Line* 39, no. 2: 160–161.

Herzog, Fred, Claudia Gochmann, Douglas Coupland, Jeff Wall and Sarah Milroy. 2011. *Herzog: Photographs*. British Columbia: Douglas & McIntyre.

McElroy, Justin. 2019. "The biggest change in the Downtown Eastside isn't the crime or homelessness. It's the geography." *CBC News*, August 21, 2019. cbc.ca/news/canada/british-columbia/dtes-vancouver-statistics-anecdotes-1.5253897.

Milroy, Sarah, and Fred Herzog (2011). "Fred Herzog: Words and pictures." *Queen's Quarterly*, 118, 2.

Muñoz, José Esteban. 2009. *Cruising Utopia: The then and there of queer futurity*. New York: New York University Press.

O'Donoghue, Dónal. 2019. *Learning to live in boys' schools: Art-led understandings of masculinities.* New York: Routledge.

Parker, Brenda and Rachel Weber. 2013. "Second-hand spaces: Restructuring retail geographies in an era of e-commerce." *Urban Geography* 34, no. 8: 1096–1118.

Probyn, Elspeth. 1996. *Outside belongings.* New York: Routledge.

Sennett, Richard. 2018. *Building and dwelling: Ethics for the city.* London: Penguin Books.

Sorensen, Katelyn and Jenifer Jorgensen. 2019. "Millennial perceptions of fast fashion and secondhand clothing: An exploration of clothing preferences using Q methodology." *Social Science* 8, no. 9, 244. doi: 10.3390/socsci8090244.

Stewart, Kathleen. 1996. *Space on the side of the road: Cultural poetics in an "other" America.* New Jersey: Princeton University Press.

Stewart, Kathleen. 2007. *Ordinary affects.* North Carolina: Duke University Press.

Stewart, Kathleen. 2008. Weak theory in an unfinished world. *Journal of Folklore Research* 45, no. 1: 71–82.

Turunen, Linda Lisa Maria and Hanna Leipämaa-Leskinen. 2015. "Pre-loved luxury: Identifying the meanings of second-hand luxury possessions." *Journal of Product & Brand Management* 24, no. 1: 57–65.

4 Black life and aesthetic sociality in the Subúrbio Ferroviário de Salvador, Bahia

Brais Estévez

4.1 Introduction

This chapter discusses the concept of Black life in the city in an attempt to explore what I call fugitive ways of re-materializing and reimagining the urban peripheries. Based on two years of urban ethnography in the city of Salvador, Bahia (Brazil), the chapter centres on the analysis of art-based initiatives by the Acervo da Laje (The Rooftop Collection, henceforth referred to as the Laje Collection or simply Laje), a self-managed museum located in the Subúrbio Ferroviário de Salvador, one of the main Black outlying areas of the city. In particular, I focus on how Laje's politico-aesthetic practices reanimate a different sense and materiality of Black life that allows Laje's inhabitants, along with its visitors and participants, to sense, think and enact Black life in the urban peripheries otherwise—beyond images of placelessness, dispossession, violence and death.

To this end, Black life is explored here through an ontology of fugitivity. The concept of fugitivity stems from the Black radical tradition and the emancipatory practices of enslaved people (Hartman 1997). Fugitivity suggests that in situations of abjection, violence and subjugation, there are always forms of Black life that escape capture and develop practices of refusal (Hartman 1997; Moten 2003; Campt 2017). However, before going on with this conceptual focus, in this introduction I aim to elaborate how the experience of living in Salvador and being immersed in the socio-material spaces of the Subúrbio and Laje, pushed me to reorient my theoretical perspective towards doing urban ethnography.

In December 2016, fresh from a Barcelona still in its post-15M Movement[1] period, I landed directly in the capital of the north-eastern Brazilian state of Bahia to conduct the ethnographic research for my project "For a cosmopolitics of urban government. Citizen labs and other urban assemblages".[2] I had designed this project in the light of discussions I had participated in my previous work in Barcelona (Estévez 2019), mainly around the challenges that both the 15M movement and the crisis of representation in Spain had thrown up in the field of urban studies. By crisis of representation, I mean the deep and generalized citizen distrust in both institutional forms of politics and expert knowledge (see Callon et al. 2011), which spread in Spain after the 2008 financial crisis. Central to these challenges was the irruption of a host of radical citizen initiatives that—without waiting for permission from city's institutional governance—were

DOI: 10.4324/9781003027966-5

intervening in redesigning the urban space, while at the same time exploring the possibilities of public space as a site for "political discussion and civic engagement" (Domínguez Rubio and Fogué 2013, 1040).

First, these urban practices resonated with the work of Jacques Rancière (2004), particularly with two concepts that lie at the core of his thought: *disagreement* and *distribution of the sensible*. Disagreement is not merely a dispute between two parties but rather a conflict which calls into question the very existence of both parties and hence of the common world based on the asymmetric distribution of the sensible. Such distribution refers in turn to the consensual order which organizes hierarchically all bodies and the whole of social space through a particular framework, a set of perceptible proofs that fixes everyone in specific positions on the basis of unequal functions, places and modes of being. Therefore, in Rancière's view politics is not about the exercise of or the struggle for power. Instead, it consists in a collective practice of dissent through which those who have no part—i.e., those condemned to invisibility, incapacity or to being heard as noise—disruptively assert their equal aptitude to see, say and make a common world. This is precisely what happened with the urban practices that I was closely following in Barcelona. These collective movements struggled to reshape the map of the sensible (Rancière 2004) that, until then, had reduced the planning of public urban spaces to a technical issue, the exclusive terrain of politicians and experts. Thus, the activists and citizens who seemed unauthorized to meddle with the design of the city introduced new ways of seeing, thinking and organizing the urban space.

Second, there was the irruption of Actor-Network Theory (ANT)-inspired approaches in urban studies, such as assemblage urbanism (McFarlane 2011) and urban cosmopolitics (Domínguez Rubio and Fogué 2017), that highly influenced empirical research in urban studies in the European context. The ANT research afforded ontological openings in the study of cities. It stressed that cities do not exist as taken-for-granted totalities but as multiple and heterogeneous assemblages, peopled by entanglements of human and more-than-human entities. Seeing the city as a multiple and decentred object enabled research that "extend[ed], interrogate[d] and speculate[d] about the kinds of things, sites, and bodies that constitute[d] the cosmos of the political" (Domínguez Rubio and Fogué 2017, 159) in the urban territory.

My intention was to use this political and theoretical experience in Barcelona to study radical urban initiatives in Salvador which were exploring new paths for the collective composition of the city. However, Salvador turned out to be very different from the city that I had imagined in Barcelona. Invited by a member of the Lugar Comum research group (n.d.) (http://www.lugarcomum.ufba.br/) at the Universidade Federal da Bahia (UFBA), I immersed myself in a collaborative network linking the university with grassroots movements, articulated around the demand for the *right to the city* (Frediani et al. 2016). This network harboured a constellation of encounters, investigations and mutual care aimed at strengthening urban struggles, giving them the visibility and status necessary to negotiate with the state. The movements involved were collective assemblages of Black urban communities (see Walker et al. 2020) that were threatened by eviction and demanded alternatives to the lack of basic facilities (sanitation, *crèches*, medical centres) while denouncing the fragility of infrastructures (housing that was on the brink of collapse). Laje was among these. More than a street protest movement, Laje consisted of a space of research and aesthetic experimentation aimed at the preservation of the artistic memory and materials of the Subúrbio. Rather than denouncing anti-Black violence and demanding improvements for the urban peripheries of Salvador, Laje's project proposed a different approach to Black geographies. One focused on the study, performativity and de-objectification of Black urban presence as not only connected to enclosure and social death but to Black life.

I understand the urban conflicts and infrastructural demands associated with Laje, and more broadly with other Black communities, in connection to Black geographies' critical effort to intersect geography with Black studies and poetics as a way to truly pay attention to mundane and minor modes of making space and place in the face of anti-Black urban policies that deny Black life (McKittrick 2011). However, the approaches I was setting out to use were essentially blind to such tense intersection between Black life and geographic studies. On the one hand, the literature of assemblage urbanism that I was studying up to this point (Farias and Bender 2010) tended to imagine non-specific urban worlds capable of achieving a more democratic organization by introducing universal ontological openings in the city. On the other hand, for Rancière (2010) politics and art constitute the generic field where humanity brings its emancipation into play around the dispute over the distribution of the sensible. However, the ontology at play in this field is still logocentric and anthropocentric (Bennett 2010), and both politics and art are seen as exclusively human and Western.

Thus, my research no longer dealt with practices of reassembling common worlds or with how a new division of the sensible would lead to emancipation. It reoriented towards the concept of Black life, thinking how it was affirmed in the city through diasporic practices and modes of existence produced within spaces of racial violence and encounter. This was so because those practices entailed a different mode of being and inhabiting the city which did not seek to be equated with settler-colonialist spatial practices (McKittrick 2013). Additionally, the research needed a sensual readjustment towards a theory of sensing Black life in the city through concepts such as *hapticality*. In the Black radical tradition, aesthetics and the sensible do not evoke a conventional idea of art nor do they imagine an aesthetic relationship determined by individual sense perception. On the contrary, this tradition considers that aesthetics and sociality always go hand in hand. Harney and Moten (2013) suggest that since the sensory history of Blackness is always connected with forms of sociality, aesthetics can be thought as a form of hapticality: "The capacity to feel through others, for others to feel through you, for you to feel them feeling you" (98). The encounter with the *Laje* Collection pushed me towards these reorientations. But before going any further in the ethnographic study, I discuss in the next section the theoretical foundations with which I reoriented the study.

4.2 Black onto-aesthetics

The Black radical tradition as characterized by the works of Hartman (1997), McKittrick (2011), Moten (2018a, 2018b), Campt (2017) and others is a philosophical and political framework around the historical and futuristic being of Blackness. It deploys an ensemble of practices of performativity and de-objectification which Moten (2018b) describes as "political, aesthetic and philosophical derangements" (141), seeking to envision Black modes of sociality and life that are not defined as the Other or the opposite of White.

While scholarship in urban studies and geography have repeatedly theorized Black life as death, and connected to countless acts of Black suffering, displacement and ruins (McKittrick 2013), the Black radical tradition suggest that Black life is a performance of fugitivity or constant escape from such representations and its deathly objectifications. The Black radical tradition is also deeply committed to witnessing and studying the sociality of Black life and its different expressive modalities.

Moten (2018a) writes that fugitivity is a "desire for and a spirit of escape and transgression of the proper and the proposed" (131). Moten does not use the notion of transgressions as invoking Western ideas of free will, moral judgement or entrepreneurship. On the contrary, the transgression that he mentions is forged "within the context of relations of domination and not external to them" (Hartman 1997, 8). These were/are relations of domination that existed and persisted within the philosophical, political and aesthetic frameworks contained in the Enlightenment as well as in the capitalist and colonial worlds constructed around the aforementioned ideas. Therefore, the Black radical tradition understands Blackness as always on the move and exceeding existing Western philosophies and politics through performative decolonizing acts.

In moving away from an understanding of Blackness as a "death-driven epiphenomenon" (Moten 2018a, 33), Black radical studies engage Black life as an infinite and unscripted range of modes of existence. This does not mean that Black radical studies see death and life as being opposite categories, but that Black life exceeds the frame of social death as much as it is continually asserted through a non-binary relationship with deathliness. In other words, Blackness requires to be approached by focusing on moments of improvisation, movement and escape from deathliness and objectification. Often, these moments are elusive, paradoxical and almost imperceptible, expressing themselves in gestures, details and modes of inhabiting which are difficult to notice and may need deep attention to identify if they are on the side of subjection or insurgency (Hartman 1997, 8). Also, the opaqueness of such modes of expression and inhabiting avoids capture and the reduction of Blackness to be a "single being" (Moten 2018a, 178).

Concerning the Black inhabitation of cities within diasporic contexts, McKittrick's (2011) concept of *Black sense of place* highlights the "process of materially and imaginatively *situating* historical and contemporary struggles against practices of domination *and* the difficult entanglements of racial encounter" (949). While conventional approaches in geography, including critical ones, tend to subsume experiences of Black people in the city in static and narrow narratives around "geographies of dead and dying communities" (954), Black urban life is more than histories of domination, suffering and resistance in "spaces of absolute otherness" (954). Consequently, a Black sense of place stresses the infinite number of place-making projects—both historical and daily—through which Black people have asserted their life in the city within racialized entanglements, that is, in the face of domination, erasure and death.

The work of the Black radical tradition around a performative ontology of Black fugitivity has also important aesthetic implications. Fugitive movements do not subordinate their courses of action to the mere recognition of Black life under the framework of formal freedoms, subjecthood and citizenship. As Moten (2018b) writes, the main specificity of Black actions refusing capture should not be sought simply in the rejection of exclusion. Rather, it is about the performative construction of an "aesthetic sociality" (160) which unfolds in the interplay of the refusal of what has been refused to Blackness. This is, it is the refusal to an "admission to the zone of abstract equivalent citizenship and subjectivity" (136). In other words, Black aesthetics does not convey a desire for inclusion in a universal common sense of freedom but affirms a form of life in common, created "from and as a sensual commune" (Moten 2008, 199) and by means of an incessant performative activity.

Such aesthetic sociality is nourished by everything that a Kantian aesthetic sees as pathological, including the worldly, the sensual, the fleshly. As such, Black aesthetics dissents from the Enlightenment project (Moten 2003). While in a Kantian aesthetic the relationship of people to art is produced through an enclosed common sense that shapes free and autonomous citizens—i.e. free from community constraints—Black aesthetic practices find their genealogy in the resistance to enslavement. They emerge in an ambiguous field between constraint and flight and refer to experiences and scenes in which "pain is alloyed with pleasure" (4). Thus, this aesthetic production entails an "affirmative refusal" (Moten 2008, 192), which both rejects the formal freedoms denied to Black people and asserts a fugitive "Black aesthetico-social life" (192).

The *undercommons* (Harney and Moten 2013) is the conceptual figure which condenses all this onto-aesthetic fugitivity. Defined as projects of fugitive planning and Black *study*, it is a political, epistemological and aesthetic idea disseminated in places, practices and projects that experiment with Blackness as another mode of life. Rather than a particular type of research, the notion of Black study refers to a radical intellectual activity that thinks of research as a form of alternative sociality conducted jointly with others in an almost clandestine way. The aim of the undercommons, according to Halberstam (2013), is not to repair things or offer clear answers to community problems, but "to take apart, dismantle, tear down the structure that, right now, limits our ability to find each other, to see beyond it and to access the places that we know lie outside its walls" (6). In the next section, I discuss how the tenets of this Black onto-aesthetics resonate in the Laje's collection.

4.3 Searching for beauty in an *unliveable* place

The Subúrbio Ferroviário is made up of several Black working-class neighbourhoods on the north-western outskirts of Salvador. In terms of reputation, this urban area has been systematically condemned by the gaze of the city's elites and corporate media to the status of an unliveable place: violent, impoverished and dangerous. Both the people who inhabit these neighbourhoods and their ways of life are stereotyped as less-than-human and expendable. Such anti-Black narrative draws on a geographic language of racial condemnation that historically identifies Black geographies as places "incongruous with humanness" (McKittrick 2013, 6).

Counter to these narratives, the Acervo da Laje is a space for the curation of and research into the aesthetic and artistic memory of the Subúrbio Ferroviário. It is currently the only museum or exhibition space existing in the urban peripheries of Salvador. It was set up by a couple, educators José Eduardo Ferreira Santos and Vilma Santos, both born and raised in the area. The project consists of two physical spaces, very close to each other, known as House 1 and House 2. Both combine the functions of a dwelling, museum and school. This heterogeneity gives rise to an incessant liveliness.

House 1, first, was built by José Eduardo's father himself on a landfill site in a sheltered inlet of Todos-os-Santos Bay known as São João do Cabrito. There, on a *laje*[3] roofed by an asbestos sheet, José Eduardo and Vilma spent the first years of their life as a couple. Since 2011 the *laje* has become an independent self-managed museum, an archive for the curation and preservation of local artworks and historical artefacts. Moreover, although this experience goes beyond the scope of this text, it is also an informal school where Vilma gives extra classes to support the local children. House 2 is the home that Vilma and José Eduardo have lived in independently since 2015. Although it was designed by the architect Federico Calabrese (2019), a member of the collaborative network mentioned above and a friend of the couple, the house echoes the modes and aesthetics of auto-construction (Holston 1991; Corsín Jiménez 2017). Neighbours and friends built it together. It was designed as an unfinished and incomplete architectural project, structurally and formally open and in constant flux. It was also conceived as both a residence and a multipurpose space for receiving people and staging encounters and activities—workshops, talks, debates and artist residences—that House 1 did not allow due to lack of space.

Laje has its origins in a series of ethnographic studies on the Subúrbio Ferroviário conducted by José Eduardo, in which three major phases can be identified. The first centred on the genocide of the area's Black youth (Santos 2010):

> I started to give classes here in the schools of my neighbourhood, and then suddenly my students started getting murdered. And I wanted to understand this and react to it. I was very concerned, and I saw myself as powerless to intervene in any significant way through my classes. It was as if the environment, the context we were experiencing, was stronger than us. And that led

Figure 4.1 Two perspectives of São João do Cabrito from House 2. Photos by the author.

me to do the Master's, as a way of channelling the research I wanted to do. The study was my adult response, deeply felt and committed and systematic, to this violence. I enrolled in Psychology because I thought it was necessary to explore the different events and paths of an adolescent's life. I finished the Master and was lucky enough to be invited to do a PhD in Public Health in which I studied the consequences of the murders amongst the youth. Because it's incredible, in Salvador an enormous amount of young people are dying and no one says anything; it's something that's not spoken about, but obviously apart from the brutality of the deaths themselves it's producing all kinds of wounds and traumas. The periphery's Black youth are not achieving an existence because they're being killed and I wanted to study the consequences of this for the youth themselves, but also for the mothers, families and the whole community.

(José Eduardo Ferreira Santos, interview by author, January 14, 2018)

A second phase investigated the systems of knowledge and geographical imaginaries reducing the complexity of the Black urban peripheries to stereotyped spaces without history, tradition, place, life or future. Finally, a third phase has developed around José Eduardo's research into "the invisible art of the workers in beauty of the peripheries of Salvador" (Santos 2014, 149). In connection to this, a turn towards Black life began to take shape in José Eduardo's project. Working with the Italian photographer Marco Illuminati, José Eduardo embarked on the search for artworks and expressions of beauty in the area. This search unearthed a host of artists of which he had been completely unaware and whom he called "invisible artists" (221). While documenting these artists, José Eduardo took a further step and began to acquire some of their works:

For us it is really important to acquire the artworks, because apart from researching the invisibility of their creators we think that it is essential to show and preserve the materiality of their work. When I began to think about the Acervo it seemed to me that a physical place was needed to record this genealogy of the beauty of the Subúrbio which spoke of us outside the stigma of poverty and violence. Through these works we could offer a meeting place between the inhabitants of the Subúrbio and these artworks which our own world had produced without us ever hearing about it. The residents of the Subúrbio did not believe that there were artists here. Because these artworks had never stayed here, they'd never been shown here, they'd always gone to houses in the centre, they'd been sold to tourists, etc., and the people of the Subúrbio hadn't been able to get to know them and enjoy them. And also, having the artworks enables us to show and say to *the city* that the Subúrbio was something more than what they said it was, that of course it existed in the city and it existed also as a place of artistic expression. Which had a memory, that this memory included artistic development and that this process continues in the present, giving birth to the future. And positioning ourselves in this way, through a dialogue among equals, making the Subúrbio visible from the historic, artistic and aesthetic points of view, we're erasing these frontiers and these hierarchies. It's a way of positioning the Subúrbio in the city and the world. Do not define us through lack! Do not define us through reduction! And, what's more, do not define us anymore because from now on we'll be the ones who define ourselves and write our own stories.

(José Eduardo Ferreira Santos, interview by author, January 14, 2018).

From 2011 the gradual accumulation of hundreds of artefacts and artworks on the Laje gave form to a collection which, shortly, would no longer be simply José Eduardo and Vilma's. The residence that until that moment had been their family house became an open space for art and memory, a project aimed at "giving back the memory and beauty of a place that has been forgotten by a succession of public powers, enabling the people to recognise what has always been denied them" (Santos 2014, 155). In other words, without giving up its residential function, that house itself would become a work of art. First, it is a place to honour the collective history of Black aesthetics memory within a neighbourhood stereotyped as lacking in memory and lifeless. Second, it is a project committed to arts-based research and the curation of artistic practices as a way to collectively explore and enact more liveable worlds.

Gradually, the Laje Collection developed a working method that I call the *Laje method*, based on the following principles, which I state below but fully develop and apply in the final section of the chapter in connection to scenes and data from my ethnographic process:

Figure 4.2 Main room in House 1. Photo by the author.

(1) Black study: Jose Eduardo and Vilma's art-based projects always began as urban ethnographies committed to the study, reanimation and re-materialization of deprived urban peripheries' wealth;

(2) Black inhabitability: the formation of the Laje Collection involved an experimentation with fugitive modes of (co)inhabitation. These found inspiration in the art of inhabiting condensed in the *lajes* of auto-constructed houses. Additionally, the re-materialization of the Laje architecture through the creation of the collection and the propagation of exhibitions in other *lajes* became the field where these fugitive modes grew and were collectively felt;

(3) Black *thingliness*: the curation of an ensemble of artworks and historical artefacts understood through their thingliness. Thingliness and thing are terms used by Moten (2018b) which point at a performative space situated at the micro level of flesh and matter noting minor vibrations, expressions, articulations that resists objectifying practices and their inherent gestures of fixing, categorizing and immobilizing.

(4) Black sensuousness: favouring encounters from the standpoint of and evolving towards a community of sensual enjoyment where the sensory is not exclusively dominated by the purely visual.

The encounter with Acervo da Laje brought the research to focus on practices of re-materialization of the urban periphery through a theory of sensing Black Life. It is important to note though that the Laje in itself was already an arts-based research project of collecting and building publics around Black art as a way of affirming beauty and life in a negativized urban context. My study sought to learn from it, keeping a discreet position, and cultivating an evolving relationship with José Eduardo and Vilma. In some ways, this meant I no longer had a research project of my own, but rather one possessed by the Acervo da Laje. In other words, what was at stake in my research was to let it learn from and be inhabited by Laje: Its initiatives, modes, working methods and modalities of encounters, in a quest for decolonialized reorientations of traditional urban studies research.

4.4 Towards an aesthetics of sociality

From the moment I realized that my research would be oriented to producing some kind of resonance in urban studies on the basis of the Laje Collection's arts-based research, I began to spend more time there for (a) visiting the House 1 collection repeatedly, often joining the guided tours; (b) attending the House 2 activities and meetings (workshops, debates, an exhibition of audiovisual projects, etc.); (c) on many occasions going there simply to talk to or interview Vilma and José Eduardo. Also, from April to August 2018, the Ocupa Lajes project took place, replicating the Laje method in four peripheral neighbourhoods of Salvador. During these four months, I was able to document activities, studies, debates and art training workshops bringing together local residents of all ages. Each of these four initiatives culminated in a major exhibition held on a *laje* in a different neighbourhood of the Subúrbio and showing the materials produced in the workshops, alongside pieces contributed by local artists. But above all the exhibition involved curating objects of material memory belonging to the family on whose *laje* the exhibition was staged.

Figure 4.3 Improvised samba at the closing time of Ocupa Lajes exhibition held in Itapuã
        neighbourhood. Photo by the author.

4.5 A collection to re-materialize the Subúrbio

It was 26 March 2017 when I first set foot in House 1. I had been warned at
the University to be careful as the Subúrbio was not a safe area. Walking the
streets, I saw many walls painted with funereal messages commemorating young
people slain in the drug trafficking wars. But at the same time, the main street was
crowded and bustling with commercial activity. Accompanied by three young
architecture professors from the UFBA, who insisted that I should know the Laje,
I arrived at the door and Vilma and José Eduardo quickly appeared, welcoming
us with hospitality, warmth and energy that I was unused to—a foretaste of the
sensuous atmosphere characterizing the Laje. They opened the *grade* (grid) of
the *beco* (narrow alley) giving access to their house and at the door I saw a handmade sign, an example of naïve art by the neighbourhood artist Perinho Santana, naming the space. Crossing the threshold, I saw private domestic scenes in
a dark area: A kitchen with a steaming pot and a living room where members of
José Eduardo's family were resting. In the centre, a steep stairway led to another
sensuous world. On the *laje*, huge numbers of sacred and profane, old and new
objects were piled up: paintings by a large range of artists; all kinds of images
and engravings; sculptures in ceramics, aluminium and wood; many *carrancas* (fig-
ureheads) and masks from the imaginary worlds of the African and Amerindian
religions; ceramics; manuscripts; sketches; an excellent press archive on the Subúrbio; cassettes with hours of interviews to local residents recorded during José
Eduardo's research projects; books on the periphery; rare books; newspapers and

magazines; wooden toys; signs and hand-painted street placards—in which the typical popular typographies of Northwest Brazil abounded—and signs local to the area—antique, defunct or at risk of extinction—rescuing the aesthetic and sensory memory of the periphery from oblivion. There were also bricks, produced by the area's old factories which brought the neighbourhood's working-class memory back to life; hydraulic floor tiles and decorated wall tiling from the Portuguese tradition (merged in Bahia) found on the beach or amongst the rubbish; shells and snails picked up on the beaches, recalling both the beauty of Bahia and the Subúrbio's seafood fishing tradition; seeds from fruit trees and medicinal plants, coal-fired irons like those which so many Black women had used to press their white employers' clothes; old kerosene lamps made by recycling empty medicine containers, food and drink cans; and found pieces of old porcelain. Vilma and José Eduardo presented these pieces to us in a wealth of detail but particularly with great enjoyment (punctuated with outbursts of laughter), infusing the visit with an infectious sensuality. One of the first things they told us was that, contrary to the usual, sight-based museum rules, "You can touch everything!".

Figure 4.4 Handmade sign made by local artist Perinho Santana at the door in House 1. Photo by the author.

The space gathered together a vast collection of items promiscuously mingled. Yet this mixture of artistic, historic artefacts and natural objects did not produce conflict. Almost any artefact seemed to fit into any category, exhibiting a certain resistance to capture by normative modes of visibility. Everything could be handled and all the objects were arranged in a disorderly freedom. On one of the guided tours, I recorded how this very freedom made some visitors uneasy since they felt the collection and its modes of visibility as wild, excessive and chaotic. Indeed, it was not difficult to link the space to the imaginaries of *undercommons* as evoking "a wild place that continuously produces its own unregulated wildness" (Halberstam 2013, 7).

Figure 4.5 A virgin and child print by artist Cláudio Pastro in front of two sacred and profane altars. Photo by the author.

I made the most of my Laje visits to discuss a range of topics with Vilma and José Eduardo. First, their curation criteria:

> Our idea is to rescue objects that contain histories, because in the Subúrbio we've lived too long without history or simply with a history that we ourselves haven't told. Art is a basic part of this history because it shows that there has always been power here, creation and life. We're sick of these narratives about the "vulnerable population" whose effect is really just to make

us vulnerable. Apart from cataloguing, and as you can see there are millions of things to catalogue, for me the most important thing is to collect, bring together the works, objects and things, not allow them to be lost, because here it's very easy to lose memory, it happens every day, and at the Laje we want to reconstruct this memory that's been snatched away.

(José Eduardo Ferreira Santos, interview by author, March 26, 2017)

Second, on the sensory proposition of being able to touch the things:

Every object is a revelation. We think that being able to touch them is important. Touch, smell, move things that are close to life. We believe in the power of sensory experience and encounters between people and things. This *Do not touch* in conventional museums already closes down possibilities. And we're committed to inventing possibilities and displacements. So touching creates possibilities that are different from simply seeing.

(José Eduardo Ferreira Santos, interview by author, March 26, 2017)

Both the ontological fluidity of the artefacts and the practice of touching them (and José Eduardo's infectious laughter) echo the freedom drive that the Black radical tradition links to Black aesthetics (Moten 2018b). The collection does not attempt to gather aesthetic objects and offer them to public view in order for the viewers to polish their taste and become model citizens. Rather it is a question of *animaterializing* the Subúrbio: Rejecting its reduction to a place of placelessness and death and inventing alternative forms of life-in-common in the afterlife. In order to do this work of re-materializing, Vilma and José Eduardo have opted for art. In other words, for performative actions and the kind of aesthetic sociality that rest at the core of Black refusal: "Here in the periphery of Salvador, art is a form of confrontation, it is like a site for us to gather better" (Vilma Santos in discussion with the author, Jun 3, 2018).

But, why is the Laje Collection such a different space from conventional museums, both in terms of curation and sensory experience? The reduction of Black people to mere things—units of cargo—in the transatlantic slave trade represents an incalculable abjection and a debt that cannot be paid off (Moten 2018b). However, this abjection also gives rise to new abilities and a new feel, by means of "a thingly resistance to the status of mere things" (9). Moten connects here to Heidegger's essay "Das Ding" (the thing) to claim the German etymology for the word *thing*. In old German, thing refers to a "gathering... of a matter under discussion, a contested matter" (145). Moten is interested in the idea that thing is not fully resolved and involves contestation. In this way, with the concepts of *thing* and *thingliness*, Moten claims for a Black politics and aesthetics of meeting the thing *qua* thing "as project and as a problem" (147). He suggests to think the thing always through the double movement of fugitivity: "from the position of the captive and thereby to enact possibilities of escape" (36) and as "a resistance to signification" (9).

I also suggest that the materials curated in the Laje can be thought more as things—fugitive and resistant to be reduced to the mere value of the sign—than as enclosed and disciplined objects. Besides, Laje's visitors and participants may be theorized less as autonomous subjects and more as *autodispossessed* members of a fugitive community.

On the one hand, this community expresses itself through collective acts of improvisation, and on the other hand, it elaborates a new feeling—neither individual nor collective—in common: Hapticality. Rather than a relationship between subjects and objects, the process of re-materialization brought about by the Laje Collection involves and evokes a relationship of a different type: a non-binary and more-than-human sensual commune committed to the preservation of Blackness and therefore "focused and arrayed against the political aesthetics of enclosed common sense" (Moten 2018b, 167).

4.6 Inhabiting as a more than one

As noted earlier, the activities of the Laje are not limited to the curation of the collection that gave birth to the project. However, it is not in the scope of this chapter to analyse in detail the activities carried out, the specific lessons learned, the materials produced and the changes that all this has achieved in the Subúrbio Ferroviário and its relationship with the rest of the city.

In this final section, I focus on an operation of disrupting the architecture that conditioned the existence and development of many other activities connected to the collection: Vilma and José Eduardo's house, which is the house too of the Collection. My interest is in how its modes of inhabitation show a radical disposition towards Black life in the urban peripheries and how this life keeps moving and proliferating under the enclosure of the formal city. The same modes of disrupting and inhabiting the architecture that turned this auto-constructed house into an archive/collection enabled me to theorize the house and the gesture itself as a divergent mode of urbanization. It is important to note that the transformation of the House 1 in a collection that Laje initiated is particularly important because it eventually propagated to other rooftops in the periphery of Salvador, which turned into occasional exhibition spaces.

Acervo da Laje emerges as a fugitive monument: An apparatus that brings together, exhibits and summons everything that flows under the city—in the Subúrbio—and that in some way also infiltrates the formal city, since urban outskirts are "the city's underconceptual, undercommunal underground" (Moten 2017, 187). Through these modes of inhabiting combining performance and architecture, the Laje Collection and Laje's method make it possible "to think and inhabit an architecture whose rematerialization makes it an architecture outside architecture" (191). From the standpoint of this outside architecture, the Black urban peripheries of Salvador are not seen simply as an urban world exposed to death and enclosed in poverty. They may well be so, but they are also "before that, in the double sense of before, the thing that underlines and surrounds enclosure" (191).

What, then, is this (under)communal world that is "against and outside and before the city" (191)? What becomes clear in the Acervo da Laje's enactments and its improvised procedures in general is the action of the undercommons making and bringing together "the real assembly that is present outside and underneath the city's absence" (190). This real assembly is not that of the sovereign state or of citizenship but an alternative politico-aesthetic imagination that unfolds a performative relationship to place at the limits of the city, formally understood. That is, counter to the common sense of the state and the measures of enclosure, capture and normalization that it deploys from and around the formal city, the Acervo da Laje articulates an extremely specific and concrete claim by means of this real assembly. Not only is there nothing bad in Black social life, but all that the sovereign power sees as pathological, as something to be corrected—i.e. already existing Black life—is and will be the sensible material for the aesthetic foundation of a new life in common.

All of this leads invariably to a twofold movement. On the one hand, the Laje as a collection strengthens and multiplies the repertoires of Acervo da Laje fugitive sociality. On the other hand, it dismantles what separates them, which is all the barriers limiting the potential for collective modes of Black life that necessarily require meeting, mixing and hybridizing. This twofold movement expresses the disruptive and antagonistic power of the Laje with the greatest clarity. Far from being a purely celebratory movement, the decolonialized curation of the Acervo da Laje embodies a fundamental litigation with Salvador. Its collections, actions and encounters do not seek to be admitted into the city but rather to exert a bifurcation that creates a new world in the city.

Notes

1 The *indignados* or 15M (May 15th) Movement was a grassroots anti-austerity movement that emerged in 2011 in the wake of the global financial crisis. It demanded a radical change of the Spanish democracy through the citizen occupations of cities' public squares (the so-called *camps*). In this sense, the 15M movement was not merely aimed at protesting against austerity policies but also enacted a new political climate (Fernández Savater 2012) under which communitarian ways of thinking, making and inhabiting the city became possible.

2 This project was funded by CAPES' National Program for Post-Doctoral (PNPD) studies. CAPES is the Brazilian Federal Agency for Support and Assessment of Postgraduate Education. I am deeply grateful to the staff from the Postgraduate Program in Geography at the Universidade Federal da Bahia. In particular, I am greatly indebted to the colleagues Pedro de Almeida Vasconcelos and Antonio Angelo Fonseca for their indispensable support. Also, I want to extend a word of thanks to Glória Cecília Figueiredo and Tomás Sánchez Criado for continuous debate and dialogue.

3 The *laje* is the flat reinforced concrete slab that forms the roof of millions of Brazilian auto-constructed homes. Beyond its technical meaning or more pragmatic and infrastructural functions, the *laje* has become one of the main sites of socialization and leisure in the Brazilian urban peripheries and represents a prime example of Brazilian underclasses' arts of habitation.

References

Bennett, Jane. 2010. *Vibrant matter: A political ecology of things.* Durham, NC: Duke University Press.

Calabrese, Federico. 2019. "Acervo da Laje. Arqueología del presente en Suburbio Ferroviario." *Rita: Revista Indexada de Textos Académicos* 12: 48–51.

Callon, Michel, Pierre Lascoumes and Yannick Barthe. 2011. *Acting in an uncertain world: An essay on technical democracy.* Cambride, MA: MIT Press.

Campt, Tina M. 2017. *Listening to images.* Durham and London: Duke University Press.

Corsín Jiménez, Alberto. 2017. "Auto-construction redux: The city as method." *Cultural Anthropology* 32, no. 3: 450–478. https://doi.org/10.14506/ca32.3.09.

Domínguez Rubio, Fernando and Uriel Fogué. 2013. "Technifying public space and publicizing infrastructures: Exploring new urban political ecologies through the square of General Vara del Rey." *International Journal of Urban and Regional Research* 37, no 3: 1035–1052.

Domínguez Rubio, Fernando and Uriel Fogué. 2017. "Unfolding the political capacities of design." In *What is cosmopolitical design? Design, nature and the built environment,* edited by Albena Yaneva and Alejandro Zaera-Polo, 143–160. Aldershot: Ashgate.

Estévez, Brais. 2019. "Reassembling Lesseps square, rethinking Barcelona: A more-than-human approach." *International Journal of Urban and Regional Research* 43, no. 6: 1123–1147. https://doi.org/10.1111/1468-2427.12767.

Farias, Ignacio and Thomas Bender. 2010. *Urban assemblages: how actor-network theory changes urban studies.* New York, NY: Routledge.

Frediani, Alexandre Apsan, Tamlyn Monson and Ignacia Ossul Vermehren. 2016. *Collective practices and the right to the city in Salvador, Brazil: Collaborative work between MSc social development practice, the Bartlett development planning unit and Lugar Comum, Faculty of Architecture of the Universidade Federal da Bahia.* London: Development Planning Unit, The Bartlett, University College London.

Halberstam, Jack. 2013. "The wild beyond: With and for the undercommons." In Intro to *The undercommons: Fugitive planning and black study,* edited by Stefano Harney and Fred Moten, 2–12. New York: Minor Compositions.

Harney, Stefano and Moten, Fred. 2013. *The undercommons: Fugitive planning and black study.* New York: Minor Compositions.

Hartman, Saidiya V. 1997. *Scenes of subjection: Terror, slavery, and self-making in the nineteenth century America.* Oxford: Oxford University Press.

Holston, James. 1991. "Autoconstruction in working-class Brazil." *Cultural Anthropology* 6, no. 4: 447–465. https://doi.org/10.1525/can.1991.6.4.02a00020.

Lugar Comum, Grupo de Pesquisa. n.d. LUGAR COMUM. Accessed June 10, 2020. http://www.lugarcomum.ufba.br/.

McFarlane, Colin. 2011. "The city as assemblage: Dwelling and urban space." *Environment and Planning D: Society and Space* 29, no. 4: 649–671. https://doi.org/10.1068/d4710.

McKittrick, Katherine. 2011. "On plantations, prisons, and a black sense of place." *Social & Cultural Geography* 12, no. 8: 947–963. https://doi.org/10.1080/14649365.2011.624280.

McKittrick, Katherine. 2013. "Plantation futures." *Small Axe: A Caribbean Journal of Criticism* 17, no. 3: 1–15. https://doi.org/10.1215/07990537-2378892.

Moten, Fred. 2003. *In the break: The aesthetics of the black radical tradition.* Minneapolis: University of Minnesota Press.

Moten, Fred. 2008. "The case of blackness." *Criticism* 50, no. 2: 177–218.

Moten, Fred. 2017. *Black and blur.* Durham and London: Duke University Press.

Moten, Fred. 2018a. *Stolen life.* Durham and London: Duke University Press.

Moten, Fred. 2018b. *The universal machine.* Durham and London: Duke University Press.

Rancière, Jacques. 2004. *The politics of aesthetics: The distribution of the sensible.* New York: Continuum.

Rancière, Jacques. 2010. *Dissensus: On politics and aesthetics.* London and New York: Continuum.

Santos, José Eduardo Ferreira. 2010. *Cuidado com o Vão: Repercussões do Homicídio entre Jovens de Periferia.* Salvador, Bahia: EDUFBA.

Santos, José Eduardo Ferreira. 2014. *Acervo da Laje: Memória Estética do Subúrbio Ferroviário de Salvador, Bahia.* São Paulo: Scortecci.

Fernández Savater, Amador. 2012. "El 15-M y la Crisis de da Cultura Consensual en España." *Periférica Internacional. Revista para el Análisis de la Cultura y el Territorio* 13: 63–71. https://doi.org/10.25267/Periferica.2012.i13.0

Walker, Julian, Ilinca Diaconescu and Marcos Bau Carvalho. 2020. *Urban claims and the right to the city: Grassroots perspectives from Salvador da Bahia and London.* London: UCL Press.

Part II

Aesthetic Practices

5 Lively pathways

Finding the aesthetic in everyday practice

Valerie Triggs, Michele Sorensen and Rita L. Irwin

5.1 Introduction

This chapter offers discussion of a pedagogical assignment used in a variety of university postsecondary education curriculum courses. In a framework of a/r/tographic research (described in another section of the chapter), the assignment requires students to maintain a weekly artistic practice throughout the university term by sharing photographs or other artistic responses of attending and responding to what the city reveals in their regular pathways. Students also write brief field notes consisting largely of descriptive and detailed observations. The assignment plays out in students' usual routes between university and home, or elsewhere in the city, or even indoors as they perform routine tasks, where they are asked to be open to noticing what they have not previously noticed, despite travelling routes or performing tasks, multiple times. At the end of the term, students share what emerges as significant for them in the assignment. Some student images and quotes are included in this chapter. Three university professors who have used this assignment in very different undergraduate and graduate courses investigate its potential for pedagogy that might promote increased or intensified experience of being in relation to realities not immediately available or accessible in current routines.

The authors consider Timothy Morton's concept of agrilogistics as a way to understand current influence on social and pedagogical practice. Agrilogistics is a pervasive attitude in which humans think they are simply "manipulating other entities… in a vacuum" (Morton, in Blasdel 2017, n.p.) while imagining the concept of nature or wild, as something outside the human. While effects are usually considered to come after causes in this widespread thinking, Morton emphasizes the need to divest oneself of bifurcations between things and consider instead, the reality of effects as contemporary to every movement. His thinking emphasizes the interconnectedness and interdependence of all things both living and nonliving, with their infinite connections and infinitesimal differences. Every shift adds to reality, and although bodies are sometimes desensitized to an awareness of it, they are never separate from this generativity.

The authors' hope in this assignment is for students to *fall into step* with things that are other than current familiarities, and find themselves in the midst of unforeclosed and unanticipated circumstances without predetermination of form or shape. The intent is for experience in the aesthetic dimension of reality, not

DOI: 10.4324/9781003027966-7

understood as new appreciation of something but as embodied events of attunement to the city through which artistic and pedagogic inspirations result. Both attunement and the realm of the aesthetic are examined in this chapter.

5.2 The city

For two of the authors, this assignment is undertaken in Regina, a city in southern Saskatchewan. On a global scale, it is considered a small city, situated amongst grids of both family farms and large corporate farming in the "grain belt" of Canada. Most postsecondary students come from small rural communities or farms near Regina. Agrilogistics is often a taken-for-granted approach to rural life especially where there is large-scale land acquisition. Even for those who have grown up in the city, agrilogistics is a pervading culture on the Canadian plains involving "massively accelerated agriculture" (Morton 2015, IV) that is a "technical, planned, perfectly logical approach to built space", which "proceeds without stepping back and rethinking the logic" (Morton 2015, IV). Sensory experience is thought to indicate direct access to reality and in this recurrent understanding of unbroken certainty between appearing and knowing, all things come to be considered as behaving in mechanical ways. In purely mechanical functions, however, only certain things matter or inform—those that pertain to certain expectations and outcomes. One sign of a result this thinking is evident in Regina's original name Pile of Bones,[1] assigned because of the massive piles of bones left on the location after the traditional buffalo population was decimated at the turn of the twentieth century, largely due to habitat loss from colonial ranching and farming. In rural areas, pervasive monocultures producing many kilometres of single crops, plants or livestock have fundamentally altered the landscape in ways that lead to mass extinction of species as well as quicker build-up of disease and pests.

The third author works with graduate students in the mid-sized cosmopolitan port city of Vancouver on Canada's west coast. Many of these students arrive from international destinations and are unfamiliar with the new city. They often find it difficult to develop a sense of intimacy with local urban terrain when they feel such strong attachment to their traditional home places. While the ocean and mountains play a larger role in the atmosphere of Vancouver, it too is profoundly located within global agrilogistic thinking, which interprets the world based on capital, resources and market. Vancouver and its metro municipalities are designed around goods and services as well as the movement of people from one place to another. Living amongst this repetition generates certain expectations; it tightens the communicative fabric that binds society together in a necessary sense of consensual truth of information. However, without the sensation of potential in what is not yet known, or impossible to know, society gets caught in a potentially implosive homogenizing feedback loop (Llinas 2001). Building on the model of land as a resource for consumption, the human is increasingly positioned as consumer and model of consumption. Children are "constructed as future producers in the global labour market via standardized assessment and curriculum... efforts to ensure accelerated preparation into the global market have affected assessment and pedagogy practices in schools" (Iannacci 2018, 220–221).

In his critique of agrilogistics, Morton is not advocating for dismantling of actual agricultural forms (Morton 2018) but rather that there is no point in trying to think outside this pervasive logistics. It is impossible to stand back from what one is entangled within. There is no detached or neutral point of observation from which to exercise the certainty involved in agrilogistic causality. Because "human sociality is always and everywhere associated with its surrounds" Thrift 2014, 4), Morton (2018) emphasizes the urgency for becoming more intimate with other lifeforms with radically different temporalities. Morton maintains that to think differently one must look for the self-contradiction in things, which he claims is an ontological wildness (in Blasdel 2017) in everything. To respond to these exhortations, one must notice how wildness keeps undermining the rigid boundaries of things including one's sense of self. The educational research question is how to accentuate these contradictions, interrupt their logic and move them somewhere else in bodies that draw on new sources of vitality.

Taking on the dynamic form of one's entangled movement involves reconfiguring oneself in relation to the physicalities and sensations of places and becoming more intimate with the unknowable, with one's limits, with sensing earth forces that emerge in the grids of everyday living and experiencing the pedagogical address in sensory experience. Based on her study of the design of learning environments, Elizabeth Ellsworth and Jamie Kruse (in Smudge 2010–2022 advocate for reconfiguring ordinary activities in such a way as to practice movement at a scale that aligns with other material conditions that shape life on earth. In this chapter's assignment, the authors ask for repeated experiences of being attentive to a sense of coexistence with life, not as already known, or even fully knowable. They hope to ignite complex and uncanny connections between places, as well as what Morton (2013a) might describe as a reverberation of the recurring aspects that make up a certain unity that is the self.

In the compression and proximity of urban populations, tensions build; cities become vigorous in creating next circumstantial moments through continuous variations that characterize the infinite encounters between all forms in their resultant displacements, compositions, endings and transformations. Sensory technology starting to actively shape its environment and cooperation between data streams are examples of endless reconfiguring of cities that move without direct human perception. It is not human logic that joins the form of one moment to the next so much as continual relationally responsive, small and large rearrangements. Ellsworth (2005) finds a pedagogical force in the fluctuating bringing forth of surprise. She explains that learning never takes place in isolation; instead, "something in the world forces one to think" (55). The pedagogical address that all things carry, and the learning self, which may involve humans or not, interfuse "over and into each other" to "become more than they ever hoped for" (55). Perhaps, in the city, students may feel more room for alternative pathways and be more often presented with the limits of ability to comprehend places as resulting only from human intention or existing just for humans. In the city, students are more likely to interact with strangers and traces of other intra-action. In shifting to align more closely with what one observes, new intimacy emerges with the strangeness of things—their markings and microbes—all teeming in the midst of other relations in time and space.

5.3 The thinking

The authors use a/r/tography (Irwin 2004) as a methodology in this assignment because it is a way of attending to how practices of artmaking, researching and teaching/learning run through one another, where other signals and manifestations ignite, erupt and proliferate in offering new paths, some to be followed and most to linger as felt potential (Triggs et al. 2013). A/r/tography is a practice-based research (Sullivan 2007) of educational practice through artistic and aesthetic experience. Anita Sinner (2017) describes it as,

> A theoretical and methodological disposition that emerged in 2003 at the University of British Columbia, Canada, among a group of education researchers and graduate students working in-between the fields of curriculum studies, fine arts and the social sciences, and engaged in both the practices of artists and the theoretical perspectives circulating in education at that time. (39)

In subsequent decades, a/r/tography has provoked international recognition and use. The forward slashes in a/r/tography mark liminal spaces but may also designate how forms of action and labour lean into entanglement with other practices and possible forms of life, indicating openness to coexistence of different paths undertaken, to connections between art and text, interruptions in the unformed spacetimes of artists, researchers and teachers/learners and to the inextricable relatedness of knowing and being. It is a methodology that invites reshaping. Nicole Lee, et al. argue, "Rather than a rigid set of methods or a set methodology to be adopted or applied, a/r/tography is a lived and emergent practice into which one enters without knowing a predetermined end" (Lee et al. 2019, 682). Other scholarships have additionally noted the pedagogical invitation that emerges in the midst of a/r/tographic research (Triggs and Irwin 2019) as well as the usefulness of a/r/tography in pedagogical design (see Barney 2019).

In this particular chapter, the authors share thinking from iterations of their parallel assignment in what Ellsworth (2005) might describe as an attempt at a loose mapping that draws on retrospective chronicling of situated and embodied encounters. It is an attempt to be "instrumental, performed as a way of getting somewhere else on the way to alternative understandings of pedagogy" (12–13). Ellsworth's understanding of such instrumental readings involves "ramifying" the effects for purposes of future actions as educators by "scattering thoughts and images into new and different alignments and practices" (13). Ellsworth claims that juxtaposing, complicating and creatively mating materials in ways that overlap and mix, and by looking at events from different angles and experiences and sometimes all at once, something new and different might be made of what one thinks they already know. Undertaking research that is collaborative and participatory is central to a/r/tographic inquiry. Cross-cultural exchanges of geo-specific understandings and diverse cultural integrations provide unique historical, social and cultural perspectives at the intersection of art and education.

The authors are strongly influenced by new materialist thinking in which all matter is an active force producing and productive of, social worlds, human life and experience. They consider the scholarship of Karen Barad who stresses the nonhuman aspect of agency that is not an attribute of something or someone but rather, the process of cause and effect in enactment (Barad 2007, 214). Agency is the enactment of "changes to particular practices—iterative reconfigurings of topological manifolds of spacetimematter relations—through the dynamics of intra-activity" (178). The reconfigurings are iterative appearances of matter and each carries new possibility and new potential. In intra-action, "apparatuses of bodily production are intra-acting with and mutually constituting one another" (389). A focus on this ecological ontological relationality in assemblages of individual events, entities and sets of practices presents challenges to agrilogistic thinking where humans are considered rational, self-aware, self-moving agents.

Most specifically to the authors' engagement of new materialist thinking is augmenting embodied feelings of being in the midst of intra-action that carries Ellsworth's "pedagogical force". The authors draw on Morton's descriptive tools for re-examining the chasm of intra-action, which is the dimension of the aesthetic. This chapter may prove useful in mobilizing discussion around the aesthetic dimension of reality as understood by Morton (2013a, 2013b, 2018) and essential to the very possibility of thought (Ellsworth 2005) and may be useful for researchers interested in the potential of what art can and might do. In upcoming sections of this chapter, attunement and the aesthetic are elaborated in more detail in relation to discussing the assignment in question.

5.4 The assignment

The assignment requires students to take photographs from different angles of whatever it is that reveals itself to them in the city. According to Lee Mackinnon (2017), a move towards a new materialist photography involves a "different kind of process than one limited to the photographic object" (50). Emphasis is placed on entangled relationalities and intra-actions that are functional extensions of bodies involved. Rather than disentangling photo from photographer and photographed moment, opportunity is presented for a process that "sketches an embodied and experimental approach" (Pyyry 2016). Noora Pyyry describes this experience through her design of photo-walks as creative educational encounters in which students feel a "moment of hesitation" (104), of not knowing exactly what to do—while held in the sensation of "incomplete knowings, ideas and sensations" (Ellsworth 2005, 17). Ellsworth and Kruse might describe this as pedagogically undoing oneself, perhaps making the "everyday feel less populated with our own human projection of plans and expectations—more real, more immediately emergent in the midst" (in Smudge 05.07.2020).

In new materialist photography, the photograph is not simply representation but rather, nonhuman agent that is complex, entangled, nonlocal and atemporal. Neither human nor image comes fully formed nor can anything be fully captured. The authors draw from Morton's (2012) understanding of poetry to borrow his words to similarly offer a new consideration of the photograph. The poem is commonly understood as an art form like the photograph but what might count as a form of art becomes less predictable in Morton's theorizing. In similarity to his understanding of a poem, the authors consider the photographic image as a "record of causal-aesthetic decisions" (219). This means that what emerges is not entirely controlled by human intention, not all of an appearance is readily apparent and what materializes is always slightly different from the things that propelled its appearance. Furthermore, some of its meaning is still to come in the future. In each return, the record is reactivated to release different meaning. One begins to understand any image or appearance as inexhaustible in vigorous subtleties of individual colour reflections, in atmospheres of particular vitalities and passions. Art in this thinking may even force viewers to "acknowledge that we co-exist with uncanny beings" (Morton 2012, 222).

After their photo encounter, students are asked to write descriptively. John Durham Peters (2015) describes writing as "an embodied practice", a wildly diverse mode of social intra-action, an excellent way to consider "human embedment in matter and media", as it "mixes the animate and the inanimate, flow and fixity" (263). In writing, one might experience a perceptual reliving of the initial event, where experience is folding back on itself in much the same continual way that Triggs and Irwin (2019) describe the city, producing new "frictions and coherences" (3). With every fold, overlap and wrinkle, self-contradictions produce a "kind of sentience" (3) not limited to human consciousness. These practices of feeling capacity to change and be changed guide what is possible, including what impinges on the body, redefining what it can do, and also, the authors presume—what images and writing might emerge.

The authors encourage students to linger only in description in their writing, rather than move to sentences that begin with "This reminds me of…", or "This makes me think about…". Asking students to remain in description rather than explanation or narratives of memory avoids direct links to definitive causal explanations that make an experience immediately understandable in relation to "knowledge already gotten by someone else" (Ellsworth 2005, 16). The uncanny sliding between form and signification is often rushed into explanation providing logical connection between things and their appearance. Descriptive writing tends instead, towards a deep coexisting with *what is given* rather than what is already known. For example, Morton argues that when students coexist with the fragile phrases of a poem they experience the many physical levels of its architecture. Writing something fragile and less explanatory does not fill in all the gaps and may encourage utterances to be more affected by forces of other words, writing and thought. The poem, or in this case, the photo event, the image, the descriptive writing are all fragile art forms that share in common, feelings of capacity to differ. In fact, Morton claims all life forms, including human bodies, are poems about nonlife. In poems, inconsistent elements are present, as nonlife comes in and out of phase, appearing, disappearing, reappearing. The meaning of the poem, Morton writes, can therefore only be another poem. Every revisiting of the city and its viscous openness to attunement offers another fragile invitation for new ways of experiencing the wild inexhaustibility of things.

5.5 The attunement

Remaining solely in descriptive writing requires effort. The present feels more familiar and continuous when one can establish a link to what one already knows. This tendency aligns with agrilogistic thought, rendering nonhuman entities more available, useable and knowable. It misses the feeling of lively pathways of contradiction in every relation, which Morton describes as attunement. For William Pinar (2019), attunement is an important aspect of study, where aligned subjects accent each other's alterity: Attuning is the birthing of difference. Pinar explains attunement as re-experiencing the subjective presence of another, to "its immediacy and specificity… but also to its situatedness, relatedness, including to what lies beyond not only spatially but temporally" (52). This multidimensional attunement involves "temporal singularity" (15), the continuity of the immemorial and the atmosphere, "intertwined with place, setting the tone for cognition, action and thought" (281). Pinar's attunement emphasizes its back-formation that involves withdrawal into the incorporeality of what will come.

Morton describes the opening subject as always already an attunement. Attunement just *is* the opening of things to other things. It is an experience of raw sweeping vastness in one's own inner space, an "oscillation between a subject's inner space and the object" (Morton 2013a, 93), a space of connectedness that exists regardless of human thought. Pinar (2019) claims attunement as occurring in the "midst of living" (269), even interrupting it. "It cannot be possessed or summoned" (Pinar, 261). In Ellsworth's (2019) work of fiction, teen character, Kallie, describes attunement not so much as listening or hearing but rather as a gnawing "sound-feeling" (14), a mysterious "something *else*—coming" (30), a sensation that takes her by surprise and for her, only translatable into drawings that are tangled and twisted, "flickering things—things changing directions, mixing with what is around them, turning into something else" (33). She explains: "I mean, I draw things that aren't what they used to be anymore but they aren't what they're going to be either… yet" (33).

Attunement is never one-sided and not just human-centred. In describing a certain song that undoes him, Morton (2013b) writes about an over-closeness where it is *the song that tunes to him*: "pursuing my innards, searching out the resonant frequencies of my stomach, my intestines, the pockets of gristle in my face" (30). The more than human world lures as body responds. Barad (2007) argues that this kind of "knowing is a matter of part of the world making itself intelligible to another part" (185).

5.6 The aesthetic

The realm of the aesthetic is an enfoldment that allows the incubation of varying variations. It is not a decoration added to things but rather, the experience of the phase-shift of the body that is contemporary to every move, the intra-active zone of attunement. In the aesthetic, entities are absorbed in inhabiting their fields of openness in relation to other fields of openness; it is the space of "more" arising in attunement to the "vast nonlocal mesh that floats in front of objects ontologically" (Morton 2013a, 24), where earth's dynamic "forces carry information from nonhuman entities such as global warming, wind, water, sunlight, radiation" (22). When bodies/entities come to a stop in this zone of fluctuating repositioning and reforming, they will have undergone qualitative change.

Because the aesthetic dimension of reality is where things affect one another and all entities are moved, it is both indeterminate and causal. In attuning, one finds oneself caught in an adjustment; Ellsworth might describe it as a repositioning. One must reform, actually, in order to move on. This situates causality in the process of attunement: Other things are reciprocally and relationally redesigning futures. In agrilogistic terms, causality is considered largely in terms of mechanical functioning that parcels out nonhumans as property, resource capital, exploiting "inert" matter in ensuring the dominance of humans and the demise of other forms of life (Morton 2012). Only the visual is readily recognized while atmospheres, unactualized potential, unknowability and reciprocal impacts are ignored. When this practice hums on in the background, left in its own implosive whorls of knowledge already known, attunement is practised in smaller and smaller dimensions of sensibility, cycling in dangerously homogeneous spins. Causality theories in this thinking are often preoccupied with explaining things away and demystifying situations in order to move forward with certainty.

Morton (2013, 2013b, 2018) argues that the aesthetic *is* causality, a relational atmospheric field of the thinking, moving and feeling of all entities as they affect and are affected by the appearance of other entities. In the aesthetic dimension, bodies have access to what is normally felt to be outside of themselves, to what seems ambiguous, self-contradictory and illusory. Variations are infinite at any scale, always incomplete and never fully knowable. Feelings of ungraspability and unspeakableness are what provide access to the inaccessible: Morton (2018) calls this the beauty experience. William Connolly (2011) writes that it is at the threshold of the impossible where one accesses the withdrawn, open qualities of things: Dynamic forces that are wild, spectral, haunting, irreducible, uncanny and causal.

Studying closely the movement of forces taking form, Brian Massumi (2008) describes the creation of semblances, new manifestations of reality, as "place-holder[s] in present perception, of a more to life" (10). "More to life" is the sensation of wild overflow that never stays in the categories human create, where humans are not totally in control and might not even be included. Awareness of this dimension seems to slip into the background as people move through habitual pathways with intentions of getting from one place to another. Massumi explains tendencies to emphasize instrumental-action-reaction affordances of everyday perception as providing necessary senses of stability, despite bodies' and selves' continual and equally necessary reconnection with relation, process and movement. Extending the self of any entity happens precisely in the aesthetic where stability and movement are phases of the same thing.

Art is a contradiction in agrilogistic thinking. Morton writes, "it is one tiny corner in…our social space where we allow things" (78) to aesthetically affect. He is considering art beyond the recordable, representational or symbolic, and rather, as abstraction and knowledge at their limit. Art is already a nonhuman being, a "sort of gate through which to glimpse unconditional futurality" (78) opening humans up to the nonhuman universe of which they are part. In Morton's definition of art, there is awareness of a broader social dimension, where humans are just one entity among others and never in an exterior position with objective knowledge. As Irwin observes, art is "not done, but lived" (Irwin 2004, 33) as in researching and teaching and in correlation with Barad (2007) who claims that practices of knowing and being are never isolable but always "mutually implicated" (185).

The aesthetic in new materialist practice compels radical rethinking of human relation to nonhuman world and far-reaching re-descriptions of an ontology of the nonhuman. Humans are not the only beings that access other beings in the aesthetic dimension. For example, Morton (2013a) describes aesthetic events that happen when "a saw bites into a fresh piece of plywood, or when a worm oozes out of some wet soil..." (19) observing that sentience may be just a difference in degree, rather than kind. Connolly (2011) draws from Nietzsche in claiming: "pleasure and pain are rare and scarce appearances compared with the countless stimuli that a cell or organ exercises upon another cell or organ" (64). Roller rinks, barcodes, pedestrian call buttons, textures of buildings and roadways also modulate the compression and expansion of the city's atmosphere, and they do so in ways that Ellsworth describes as offering "material correlates for the experience of the learning self in the making" (20).

5.7 The student work

In a/r/tographically analysing student writing and photographs, the authors bore in mind their intentions for unsettling agrilogistic perception through embodied experience. While students derived much meaning and significance for their own lives, the authors consider the most important aspect of this research to be practising the sound-feeling-thought that Morton and Ellsworth describe, of aesthetic listening—attuning. Therefore, in this chapter, the authors begin to respond to student writing and art by listening-attuning-thinking to where they themselves (1) feel a coexistence with *indeterminacy*: Expressions of both presence and absence, a sense of things moving through one another, affecting and being affected, where categories seem less defined and more fragile; (2) feel a coexistence of *intimacy*: with nonhumans—uneasy senses of shifting to what is not fully understood, where things feel strange, uncanny or not fully graspable.

5.7.1 Indeterminacy

Rachel Morgan is a Regina undergraduate teacher education student taking a course titled *Schooling and Sexual Identities*, in which artmaking is encouraged as a way to understand new learning. In one week, Morgan posts a photo from a roller derby game (see Figure 5.1). In it, she is smiling even though she is in the penalty box where she writes that it is common to feel guilt and shame for one's infraction that has let down the team.

She describes the moment as follows:

> Everyone around us melted away, including the photographer who happened to look our way. A fraction of the game caught on film but a snapshot

Figure 5.1. Untitled by Rachel Morgan (2018).

of all of the aspects of roller derby that I love. We slam our bodies against each other fuelled by the intoxication of… competition. Yet in the swirling of the storm, we find ourselves enjoying 30 seconds of nothing significant…

The writing and image tunes to the authors, generating a moment that seems to be bursting with parts and wholes, scales, temporalities and sexualities, moving through one another. The authors sense the intensity of the aliveness running through the image and writing, totally graspable in the openness of image and writing to other assemblages but not fully knowable. Ellsworth (2005) explains that while the "knowings" of the aesthetic are not tellable, they are sense-able, "the untapped fluidity of the world, the movements, vibrations and tranformations that occur below the threshold of perception ….and outside the relevance of our practical concerns" (163). Morgan describes this indeterminate moment as her favourite moment in the entire game.

Another experience of being in the midst is highlighted by Ken Morimoto, a Vancouver graduate student in an A/r/tography Research course (see Lee et al. 2019). Morimoto investigates movement between familiarity and unfamiliarity by inquiring how far one must walk in order to be moved, in other words, to become unfamiliar. Morimoto created a series of three five-day barcodes composed of photos taken over three series of five days—one photograph every hour he was awake. At the end of each day, he edited the photos into a barcode format.

Between combined strips of photographs of each day, Morimoto imposed black bars equivalent in size to his sleeping hours (see Figure 5.2). He draws

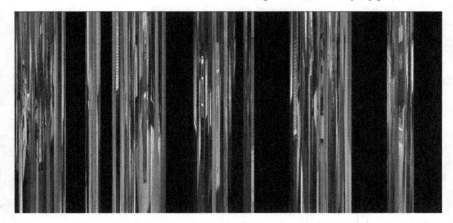

Figure 5.2. Five Days in September by Ken Morimoto (2018).

from Izutsu and Izutsu (1981) to compare black bar breaks to those in Japanese court theatre, which when mastered, make visible internal movement through external stillness between movements.

In writing, Morimoto (2018) notes the effects of surrender of each photo to the totality of the barcode as follows:

> The details that might reveal the time and place were reduced so that one can only approximate the when and where of each barely distinguishable strip by the little they see and by the information the surrounding strips provide... By weakening the visual strength of each photo... each strip becomes more active relationally. One hour begins to correspond with its surrounding counterparts, gaining certain qualities based on how they interact with one another.
>
> (13)

The black strips indicate potential for internal stillness, expressing how things emerge from within events: Presence and absence together, the inclusion of the invisible within the visible.

In the process of each strip becoming more indeterminate, each becomes more relationally activated. By limiting representational recognizability of the strips, their causal dimension starts to appear. Each strip of existence begins to make things visibly happen. Emerging encounters of the realm of the aesthetic are evidenced in Morimoto's work as he begins to sense how "an artwork is subscended by its parts" (Morton 2018) rendering wholes "less than the sum of their parts" (84). Each added strip of experience affects another. Morimoto plays with the push and pull of the fragilities of both wholes and parts where there is always the danger of one piece of existence or one whole, enveloping another in destructive ways. The authors are reminded of how their own attunements to student work are always part of opening and closing other futures.

5.7.2 Intimacy

Marzieh Mosavarzadeh is also a PhD student in a Vancouver graduate A/r/tography. Research course whose work summoned for the authors a sense of intimacy with the nonhuman world. Mosavarzadeh (2018); see also Lee et al. 2019) writes that walking in the city evokes an active and gradual formation of her own hybrid becoming, based on memories of home but conceived within an unfamiliar place. Immigrating to Vancouver is a mixture of distant and nearby places and times and continuously going back and forth between places and moments provides feelings of being suspended between two utterly different worlds.

Mosavarzadeh's photographs include a variety of objects such as pedestrian call buttons (see Figure 5.3), coin laundry marts, SkyTrain tracks, walk signals and care share systems, all unfamiliar objects that she did not encounter in her home city of Tehran.

Mosavarzadeh writes that she begins to feel a strange sense of attraction and attachment to these objects, one that is hard to explain. Attuning and documenting them offers a newly found sense of closeness to them. Morton (2013a) argues that when we tune into the more than human world, real things happen. In attuning, "we are affecting causality. We are establishing a link with at least one other actually existing entity" (23).

Mosavarzadeh also experienced a sense of intimacy with the other that was herself. After making postcards from her photos, she wrote a note on the back

Figure 5.3. Untitled by Marzieh Mosavarzadeh (2018).

relating to a certain object and sent it to herself in Tehran, to the part of her that "stayed and never moved" (Mosavarzadeh 2018). Mosavarzadeh considers her postcards portrayals of unremarkable things: A collection of the mundane. However, she writes, that to herself in Iran, the very mundaneness is what breathes life into them. To her, they are windows into a parallel universe, bits and pieces of ordinary life in a distant place, bringing her closer and closer to the her, that is in Vancouver. She writes that she keeps returning to her Tehran self as "an acknowledgement of the existence of her other self; an admittance that while Marzieh in Tehran held onto her home, Marzieh in Vancouver, held onto her too".

The assignment provided a sense of the simultaneity of presence and absence for Mosavarzadeh. She too began her research in this assignment by asking where the unfamiliar becomes familiar and through her inquiry, eventually realized their continual entanglement. The authors are compelled by Mosavarzadeh's experience of intimacy with nonhuman entities and the authors as well, began to consider previously unnoticed items in less predetermined ways. Consciously working to prevent herself from coinciding with herself and by bringing herself into intimacy with strangers, including foreign objects and her own changing self, Mosavarzadeh is able to sense a coexisting extension of her Tehran self in the city of Vancouver.

Kyla Crawford, a Regina Arts Education second-year undergraduate student, creates a haptic map of her weekly excursions (Figure 5.4). She is interested in how tactile perception affects her emotional reactions to each place. By bringing slices of plasticine modelling clay with her to the places she documented earlier with photography, she creates a map that produces the textures of each place. In each site, Crawford chose the texture, which she felt most expressed its presence

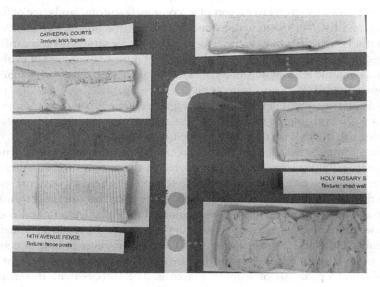

Figure 5.4. Untitled by Kyla Crawford (2019).

to her. She pressed the plasticine onto the texture to obtain a negative impression, generating new texture through attunement to texture.

Later, Crawford mapped out the images on poster board in the format of a subway system with each place charted in relative distance to each other rather than to scale. She intends viewers to see the map in its entirety but also close their eyes and feel the distinct textures guiding them. Crawford painted the path of travel with thick strokes of acrylic paint in order to create a rough traceable surface. A yellow "stop" button indicates each place and painted dots extrude so the toucher can follow the pathway from place to place, stopping along the way to feel the textures of each place.

In their work on the art of film, Laura Marks and Dana Polan (2000) explain how "haptic perception privileges the material presence of the image" (163), working at the "level of the entire body" (188), and stimulating more intense intimacy with the world, "located on the surface of the body (163) rather than with optical visuality, which only "sees things from enough distance to perceive them as distinct forms" (162). Crawford's assignment seems to play with the question of whether there even is, an outside to experience. Instead, her work offers the authors attunement to the strange wildness of being close, even too close to know places in any determinate way, and to the experience of the intimate of the aesthetic that is not fully capturable, yet accessible in the indirect streaming sensorium of the city.

In summary, while agrilogistic thinking has blocked and suppressed lively pathways between causality and the aesthetic (Morton 2015, II), an "agrilogistic retreat is [not] the only way to move across Earth" (XV). By falling into step with the realities of matter in the city that are the dynamic and shifting entanglements of relations, students experience openings and relations not previously felt. Much is dependent on the layering of sentient, haptic indexes of change in the city, each layer a part of a "constantly shifting landscape of forms" (Triggs and Irwin 2019) in which "humans are not the only actors" (7). While the authors argue that one can pedagogically design for attunement, it cannot be assured or predetermined and instead, to get a glimpse of "a process of evolution of a topographic sensing/sensibility" (11), they agree with Thrift who argues for "investing more effort into generating and investigating vague feelings, rather than direct causality" (11). Focusing on vague feelings and fragile responses generates a an a/r/tographic sensibility where students are encouraged to practice opening the worlds with which they are currently familiar.

The assignment described in this chapter aims to provide an opening to notice the aesthetic dimension of existence offered in the continual materializations/ matterings of the city, where it is possible to tap into alternative energy sources than the ones that drive agrilogistics. By holding open the assignment throughout the term, more opportunities were afforded for embodying radical coexistence with causal indeterminacy in more intimate ways. By requiring something shareable the burden of the aesthetic is ramified, however briefly, in comings and goings of other dynamic forms that are innovative spaces of refuge for conceptualizing a pedagogy of adaptive practices. The authors share this mapping of the various tracings and trajectories of their assignment as well as their own sensations of aesthetic attunement to student responses, in hopes that the chronicling will proliferate its effects for divergent and lively pathways of further educational research and pedagogy.

Note

1 Translated from the Cree word *"oskana kâ-asastêki"*. Often shortened to Wascana, it is now a name currently given to parks and streets within the city of Regina.

References

Barad, Karen. 2007. *Meeting the universe half way: Quantum physics and the entanglement of matter and meaning*. Durham, NC: Duke University Press.

Barney, Dan. 2019. "A/r/tography as a pedagogical strategy: Entering somewhere in the middle of becoming artist". *The International Journal of Art & Design Education* 38, no. 3: 618–626.

Blasdel, Alex. 2017. "A reckoning for our species: The philosopher prophet of the anthropocene". Thursday, June 15, 2017. https://www.theguardian.com/world/2017/jun/15/timothy-morton-anthropocene-philosopher

Connolly, William. 2011. *A world of becoming*. Durham, NC: Duke University Press.

Ellsworth, Elizabeth. 2005. *Places of learning: Media, architecture, pedagogy*. New York: Routledge.

Ellsworth, Elizabeth. 2019. *Solid, broken, changing*. Brooklyn, NY: Dragon Tail Books.

Iannacci, Luigi. 2018. Colonizing the desire of culturally and linguistically diverse (CLD) children in early childhood education: Curriculum and the creation of consumers. In *Contemporary studies in Canadian curriculum: Principles, portraits and practices*, edited by Darren Stanley and Kelly Young, 203-236. Canada: Brush Education.

Irwin, Rita L. 2004. "A/r/tography: A metonymic métissage". In *A/r/tography: Rendering self through arts-based living inquiry*, edited by Rita L. Irwin and Alex de Cosson, 27–38. Vancouver, BC: Pacific Educational Press.

Izutsu, Toshihiko and Izutsu, Toyo. 1981. *The theory of beauty in the classical aesthetics of Japan*. Netherlands, The Hague: Martinus Nijhoff.

Lee, Nicole, Morimoto, Ken, Mosavarzadeh, Marzieh and Irwin, Rita L. 2019. "Walking propositions: Coming to know a/r/tographically". *The International Journal of Art & Design Education* 38, no. 3: 681–690.

Llinas, Rudolfo R. 2001. *i of the vortex: From neurons to self*. Cambridge, MA: MIT Press.

Mackinnon, Lee. 2017. "Toward a materialist photography: The body of work". *Third Text*, 30, no. 3–4: 149–158.

Marks, Laura U. and Polan, D. 2000. *The skin of the film: Intercultural cinema, embodiment, and the senses*. Durham, NC: Duke University Press.

Massumi, Brian. 2008. "The thinking-feeling of what happens". *Inflexions* 1, no. 1 (How is Research-Creation? May 2008): 1–40. www.inflexions.org

Morimoto, Ken. 2018. "*Moving in barcodes: An a/r/tographic inquiry of place through time*". Unpublished course paper. The University of British Columbia, Vancouver, Canada.

Morton, Timothy. 2012. "An object-oriented defense of poetry". *New Literary History* 43, no. 2: 205–224.

Morton, Timothy. 2013a. *Realist magic: Objects, ontology, causality*. Ann Arbor, MI: University of Michigan Publishing.

Morton, Timothy. 2013b. *Hyperobjects: Philosophy and ecology after the end of the world*. Minneapolis, MN: University of Minnesota Press.

Morton, Timothy. 2015. "What is dark ecology?". *Changing Weathers.* http://www.chang-ingweathers.net/en/episodes/48/what-is-dark-ecology

Morton, Timothy. 2018. *Being ecological.* London, UK: Penguin Random.

Mosavarzadeh, Marzieh. 2018. "Navigating the unfamiliar: An a/r/tographic project". Unpublished course paper. The University of British Columbia, Vancouver, Canada.

Pinar, William. 2019. *Moving images of eternity: George grant's critique of time, teaching and technology.* Ottawa, ON: University of Ottawa Press.

Peters, John Durham. 2015. *The marvelous clouds: Towards a philosophy of elemental media.* Chicago: University of Chicago Press.

Pyyry, Noora. 2016. "Learning with the city via enchantment: Photo-walks as creative encounters". *Discourse: Studies in the Cultural Politics of Education,* 37, no. 1: 102–115.

Sinner, Anita. 2017. "Cultivating researchful dispositions: A review of a/r/tographic scholarship". *Journal of Visual Art Practice* 16, no. 1: 39–60.

Smudge Studio. 2010–2020. FOP Friends of the Pleistocene (Elizabeth Ellsworth and Jamie Kruse) Online blog: http://www.smudgestudio.org/

Sullivan, Graeme. 2007. "Research acts in practice". *Studies in Art Education* 48, no. 1: 19–35.

Triggs, Valerie and Irwin, Rita L. 2019. A/r/tography and the pedagogic invitation. In The International Encyclopedia of Art and Design Education, Volume Pedagogy, edited by Nigel Meager and Emese Hall, 1-16. Hoboken, NJ: John Wiley & Sons, Inc.

Triggs, Valerie, Irwin, Rita L. and O'Donoghue, Donal. 2013. "Following a/r/tography in practice: From possibility to potential". In *Inquiry in action: Paradigms, methodologies and perspectives in art education research.* edited by Kathy M. Miraglia and Cathy Smilan, 253–264. Reston, VA: NAEA

6 A hauntological enlivening of the Coma Cros archive through pedagogical inquiry and live performance

Judit Vidiella

6.1 Introduction

In this chapter, I address the challenges of working affectively with urban history archives. I do this in the context of a pedagogical experience with young university students enrolled in a performing arts degree in Salt (Girona, Spain). The programme was housed in a renovated tertiarized building that from 1850 to 1999 had been first a cotton mill and later a textile factory named Coma Cros. This pedagogical encounter was undertaken with 15 students in a unit titled Expressions of Contemporary Culture.

The pedagogical project took as its emplacement the building, the former cotton mill, now converted into a cultural and educational facility, to explore the following question: As performance artists and scholars, how can we experiment with modes of awareness and feeling through inhabiting a space that relives the historical and social narratives haunting it?

One of the most transformational moments in the project was when my students met a group of Coma Cros former women workers. Most of the students were young women in their 20s, many combining their degrees with part-time work. I sensed that clear parallels existed between the lives and schedules of the former Coma Cros workers and my students subjected to the university's logic of linear time and productivity. The Coma Cros workforce had essentially been made up of women whose bodies were subjected to the discipline of temporal rhythms framed in daily, weekly and annual timetables. The women had to juggle housework with double shifts at the factory. Figure 6.1 shows an old picture of a group of women workers used for a banner that is currentluy covering part of the building's façade with a passage from Hardt and Negri's Empire (2000). The picture acts as an affective force that vibrates and echoes with the presence of contemporary women students in the building, turning them into the new workers who also have to move in step with timetables, shifts and workplaces.

In my understanding, this picture and other traces of the past in the building and the area worked performatively and affectively. Conceiving the Coma Cros archive as an affective performance connects with what feminist new materialist scholar Tamboukou (2017) describes as "a pedagogy of the event from the creation of an archival assemblage" (2). This pedagogy looks at the submerged, marginalized histories of working women and interrogates the genealogical order of the archive,

DOI: 10.4324/9781003027966-8

Figure 6.1 On the left, Coma Cros workers leaving the factory (1910, Joan *Vivó* Holding). Image courtesy of Salt Municipal Archive (CAT.AMS.613.80). On the right, an image of the banner covering the Coma Cros building (2019) that contained the same historical image and a quote from Hardt and Negri's book *Empire*. Images by the author.

thereby creating other archives. Tamboukou (2016b) is an expert in archive theory who argues that archives require dynamic, embodied and process-based counter-narratives that reveal discontinuities, ruptures and multiple nodes of emergence. Additionally, she conceives the archive as traversed by multiple rhythms and open to new ideas and encounters (Tamboukou 2014). In short, the archive is seen as an apparatus of experimentation whose configuration has an impact on the type of data and the kind of knowledge that it will generate. Bringing together the experiences of past women workers and students' lives in this project was an agentic intervention in the histories of their times, capturing moments of being (Tamboukou, 2016a). That is, by inquiring into oral and written narratives as well as material objects and spaces connections emerged between the intense and intimate moments, often invisible, between work, educational and personal spaces.

The method of creating this archival assemblage did not only consist of compiling the women's memories of work, as it is typical in feminist oral history. As we encountered the Coma Cros archive and the lived recollection of the women workers, the students and I learned to attune our senses to the minor, intimate and intense moments that are often invisible in conventional archive practice. Concepts from live performance art and the theatre of objects that I commonly teach in the unit helped us to resist a literal, linear reading of the archive and to mobilize it as a vibrant aesthetic object and ecology. Eventually, the archive's haunted or silenced afterlives challenged our assumptions and reasoning about the past, bringing into being a new affective connection with it.

This inquiry into silenced and invisible stories led us to develop what Blackman (2019, 2012) describes as a *hauntological analysis*. The hauntological analysis of the archive sees *data* as haunted: Data are understood in relation to their affective ghosts. The afterlives of the archive can challenge our assumptions about rationality, consciousness, cognition, etc. According to Blackman, *haunted data* disrupt the distinction between big and small data to explore what is outside the frame of counting and measuring, but also in allusion to how certain ideas, things and entities have become marginalized, excluded or repressed in "historical continuity", or "are passed and transmitted through silences, gaps, omissions, echoes and murmurs" (Blackman 2012, xix). The project discussed in this chapter used these ideas to trace the ways in which, by bringing the archive alive, our research was able to haunt and contaminate the city, the university, the factory and the students, thereby transforming the relationships between bodies-times-spaces-atmospheres.

This pedagogical research experience culminated with a set of live performances titled *Estirant del fil*[1] (*Pulling on the Thread*), which also concluded the unit. The pieces were curated and performed by students during the Cultural Week, an event celebrated annually by the University. They not only performed the archive of Coma Cros but also what was *not* there yet resonated hauntingly, such as the women workers' recollections, murmurs of neighbours telling stories, etc., as an entanglement of past and present. This called into question an instrumental approach to the study of theater that separates learning techniques and subject matter from situated engagements with the knowledge and life of the places in which processes of study are embedded.

6.2 The recent history of Coma Cros

Coma Cros is located in the area of Salt on the outskirts of Girona, the 11th most populated city in Catalonia. A small agricultural town, Salt grew significantly in the mid-1800s with the installation of three Cotton Mills, one of them Coma Cros, with its surrounding neighbourhood, the Veïnat. The area's industrial development was made possible by the water of the River Ter and particularly the Monar Canal, which provided the hydraulic energy to move machinery.

Affordable housing and industrial development have traditionally made Salt a city where immigrants can find economical places to live. As a result, Salt is amongst the Spanish cities with the most ethnically diverse population. The international migration of recent decades has mixed with the local population, many of whom are also former migrants from other parts of Spain. On occasion this has resulted in cultural, social and racial tensions which have drawn attention to Salt's decaying urban fabric, and as a result the local administration has invested in renewal. The building of facilities representing this regeneration included El Canal theatre complex, which hosts the prestigious Temporada Alta International Theatre Festival.

Coma Cros functioned as a factory from 1850 to 1999, producing cotton-based and later polyester fabrics. In 1910, it was bought by Joan Coma Cros, who expanded the building, modified the course of the canal and built an electric power plant. In addition to its industrial activity, Coma Cros also played an important social and educational role. A nursery was opened there in 1954, today converted into a private retirement home. It also provided a venue for sports such as football, roller hockey, skating and other leisure activities. During the Spanish Civil War (1936-1939), Coma Cros counted among the 70% of Catalan factories that were collectivized, mainly by antifascist militias led by the CNT[2]-AIT. Factories, transport, public services, electricity, commerce, cafes, hotels, etc. were seized by unionized workers. The government of Catalonia (the Generalitat), which supported the Second Republic, saw this as central to its programme of social revolution within a war economy. Coma Cros was collectivized by decree in August 1936, first renamed Colectiva Obrera, then in October of the same year the name changed again to Kropotkin Colectiva Obrera Coma Cros in honour of the Russian anarchist Piotr Kropotkin.

During this period, the factory alternated between producing bed sheets and military equipment such as howitzer sheaths, grenades and cloth for parachutes and uniforms. The name *Kropotkin* was emblazoned on the main façade, but at the end of the Civil War the sign was painted over to avoid retaliation. Figure 6.2 shows a photomontage by local historian Jaume Prat for a book about Coma Cros, in which the sign on the façade is visible along with a militiaman, not present in the original picture, carrying a shell. Weathering over time made the sign reappear—like a ghost—in 2000. The façade was rehabilitated in 2004–2005. When the factory closed due to debt, the building was donated to the City Council and today it is a Culture Factory of 18.000 m² housing the Municipal Library, an Ateneu Popular (a grassroots cultural-educational association), the premises of various clubs (the Festival Commissions, for example), the head offices of several universities

Figure 6.2 Photomontage courtesy of Jaume Prat.

(face-to-face and online) and the Municipal Archive, containing the Coma Cros archive that we explored through this project.[3]

6.3 Thinking material performativity through new materialisms and performance theory

The aim of this section is to briefly situate theories of live art and performance art in connection to feminist new materialisms. Central to these three sets of theories is a questioning of objects as passive, things as inanimate, and matter as inert and waiting to be infused with life. Feminist new materialisms share with the field of performance studies (see, for example Chen 2012; Jones 2015; Schneider 2015; Braddock 2017; Harris and Holman 2019) and the theatre of objects (Schweitzer and Zerdy 2014; Larios 2018) the dismantling of binaries between the animate and the inanimate, life and death, the human and the non-human. Additionally, all three fields see the non-human world as made up of *actants* (animals, ecologies, objects, ancestors, spirits). This contributes to a different performative agenda, centred on fields of immanence that humans enter in the midst of existing, ongoing activity to think with many potential others (Schweitzer and Zerdy 2014; Ulmer 2017).

In this respect, the new materialisms theorize the *agentic capacities* of objects (Bennett 2010; Schneider 2015), displacing the prioritization of language as the primary way of thinking about meaning-making. This revitalizes matter by giving it a leading role as an agent of *iterative materialization* (Barad 2003, 826).

Artists working in the theatre of objects argue for

> a logic of *thingness* which is not… that objects have the kind of agency we decide to map onto them, but rather agency as per Latour's *actants* in which objects have an agency that may not yet be recognisable to humans.
>
> (Harris and Holman 2019, 12)

The *thingness of objects* resonates with Bennet's (2010) concept of *thing power*, which affirms the active role of materials in public life, their power to provoke effects and affects. From this perspective, objects are irreducible to anthropocentric visions in which they remain passive and are understood simply as consumables.

As I mentioned briefly in the introduction, discussions of material performativity also affected the archival research in our project, where the archive was conceived as an event rather than a static container of documents. According to Tamboukou (2008), this means that "the event is inside what occurs" (366), it is an eruption that results in discontinuities. It ruptured our modes of analysis and interpretation of the archive as we were in turn affected in this process. Ultimately, the archive may also be a force of imagination that initiates something new in the process of archival understanding (Tamboukou 2015, 156). This is similar to saying that the archive is a potential agent of *in(ter)vention*. With this impulse in mind, we visited the Salt municipal archive several times. The first visit was guided by the archivist but subsequent visits were free and the students organized them in accord with their interests in exploring specific documents that could eventually inspire the creation of their performances.

6.4 Encountering archival silence, discontinuity and intimacy

When the students and I visited the Salt municipal archive, an initial observation was that the documentation was organized with a focus on the performance of technical management (McKenzie 2001). Employee files, records of leave, payrolls, labour regulations, productivity reports, etc. made up the bulk of the existing documents. Tags and keywords included terms such as "industrial activities", "the war industry", "communications during the women machine weavers' conflict (1932 strike)", etc. With this territorialized way of organizing the archive, what I suggested to the students was to let the documents surprise and inspire wonder, to interrogate pre-existing categories and to let questions and curiosities redirect the attention and routes of interpretation. This idea is in tune with Tamboukou's (2017) argument that archive documents are formative rather than illustrative of evidence or, as she writes, "traces of the past that rather than representations that mirror it, [...are] conceived as fragmented assemblages, that [allow] us to imagine their lives before and after our encounter with them" (2).

With this approach in mind, we paid attention, as we explored the archive, to absences, gaps, silences, contradictions and places where data became "attractors" (20), such as photos, objects, events gathered together through discord, discontinuity, and temporal clashes or collisions. For example, one "collision" that created a moment of affective intensity rooted in feelings of anger, dis/belief and the desire to say something more was when the students came across a collection of photos from the centenary of Coma Cros (1950). These showed something like a party where a group of clearly objectified women appeared among mattresses and other manufactured products as part of an advertising display. The photographs inspired one of the students' performances addressing consumerism and contemporary fashion icons, in which the mattresses were used as a projection screen. The process of the *hauntological analysis* (Blackman 2012; 2019) of the Coma Cros archive led us towards creative explorations of enchanted matter.

For Blackman (2019), *hauntology* is a performative methodological approach permitting interpretations that transform the thing they interpret (18). Working with the objects of the archive affectively transformed them in the sense that they were enlivened, evoking silences, lacks and opacities. Following this twofold methodological orientation, archival objects and materials such as blankets, shirts, threads and others that were not there but would appear later—bread, chocolate, undergarments, etc.— became active agents in the scene. They brought us closer to the intimacy of the archive. We felt that many data and experiences were missing there and sought other ways of understanding the archive as an atmosphere that brought us closer to ephemeral, oral and somatic data. Embodying data by means of performance allowed us to inhabit non-linear time, or in Schneider's (2015) words, to "trade in animacy, the 'copresence' of living beings with the 'here and now' of space and time" (9). In a way, this affective mode of sensing the data, of seeking their ghosts and hauntings was a determination to restore the silences of the archive.

Weeks before the performances were given by the students in my unit, a "haunting" occurred. I attended a session taught by a colleague at the School of Fine Arts also consisting of students' live performances. I was enchanted by Paula Delito's piece entitled *Who was Amalia Alegre?* as it was linked to the shared themes of archive work and women workers that the pedagogical project in Salt explored. The performance centred on an anarchist activist involved in the women's revolt of January 17, 1918 in Barcelona. This street revolt emerged as a response to the post-WWI hunger crisis, Catalan anarchists called women to march in the streets demanding bread for their children (see Figure 6.3). I invited Paula to repeat the performance at our festival, held on January 16, 2017. It was at that moment that we realized that our Cultural Week was taking place 99 years after the original women's revolt with only the difference of one day. In the performance, Paula used bread and breadcrumbs as active agents and connectors in the scene, scattering the crumbs from a metal ladder by blowing on them, then handing out pamphlets and inviting us to join the cause in a communal bread-eating ritual.

6.5 *The class in the street*: The archive as an event

Our first visit to the archive motivated us to engage with it as an event, focusing on its embodied dimension. We aimed to trace the archive's silences and uncover its temporalities by re-enacting elements of it that were not present among the files, documents and stored artefacts. The archive as an event is an idea that connects strongly with contemporary Performance Studies. Taylor's (2002) concept of *repertoire* is central here. *Repertoire* refers to "performance functioning as vital acts of transfer" (2). It consists of ephemeral embodied knowledge such as gestures, dance, singing, etc. which trigger memory and remembrance. *Repertoires* highlights the idea that we participate in a system of learning, storage and transmission of heritage that expands what is seen as knowledge. *Repertoires* challenges logocentrism or the predominance of writing in Western epistemologies, privileging embodied and non-verbal modes of expression in our participation in and transmission of social knowledge and memory. Based on this, Larios (2018) suggests that there are a range of strategies that escape the traditional politics of archiving and documenting (301–305). These include: *Memes* (songs, stories and ways of doing things that we copy and imitate) and *mnemonic systems* or *mnemobjects* in which a word or object helps us remember something. Similarly, Tamboukou (2016b) conceives the archive as an intra-action that re-enacts subjugated knowledge, "a site for counter-memories to emerge and for *mnemonic* practices to be revealed" (85).

This desire to turn the archive into an event spurred us to invite former workers from the factory to meet and talk with us. To this end, we hosted a meeting in the street, titled *the class in the street*, where the students, women workers and I gathered. The meeting was possible thanks to the collaboration of local geriatric services, who connected us with some former workers currently living in the retirement home across the street from the factory. We all sat in a circle of

Figure 6.3 Paula Delito's performance *Who Was Amalia Alegre?* Images by the author.

chairs situated in front of Coma Cros (see Figure 6.4). The memories shared with us that day were painful and shocking. The women narrated how noise was so ear-splitting inside the factory that it caused actual deafness. The bosses, mostly men, abused their power, which sometimes included flirting with the women. The women developed forms of resistance and mutual aid. For example, when one of them was ill or pregnant the others would help them out or even cover their shift. The combination of a full workday with housework and children care was incredibly demanding and stressful. Children as young as seven worked

Figure 6.4 The class in the street (above) and an image of the performance *Metallic Soul* (below). Photos by the author.

long hours in the factory with no access to education. There were no breaks for changing and washing when women had their periods. A number of workers committed suicide and drowned themselves in ditches. Rumours of death and despair abounded and grew over the years. Listening to these stories was not only a privilege for us but also an incredible moment in the mobilization of tensions, silences and absences in the material archive of Coma Cros.

One of the workers' stories evoking silences and *mnemonic archives* is that they had invented a sign language as a secret means of communication in the factory. We asked them to perform for us the physical gestures they used in their work-places as a way of embodying this *repertoire*. These revealed a fascinating cho-reography of gestures that were repeated later in a performance, *Metal soul*, that the students and I co-created. It inventively mixed the aforementioned secret gestures of the Coma Cros workers with the Spanish Republic call for workers' freedom and with a song from the period inviting women to fight for the cause of the Spanish Republic. Additionally, the performers animated objects such as pots, pans and gas masks.

Together we studied Berstein's (2011) concept of *scriptive things*, an approach to archive objects that seeks to recover lost performance repertoires. It inquires into the way particular objects shape or direct human behaviour, in a "dance with things" (Bernstein 2009), and questions subject–object binaries by asking how a *thingcentric* methodology might allow scholars to uncover hidden repertoires of oppression and resistance. As an *archive of repertoires*, the *scriptive thing* collapses Taylor's binary division between the archive and the repertoire, offering a way of theorizing how human agency emerges through constant engagement with the stuff of our lives (in Schweitzer and Zerdy 2014, 11). What role could perfor-mance have here? Our project sought to understand the *ephemeral archive through performance* in resistance to the Western logic of accumulation and ocularcen-trism, opting for different epistemologies in which the oral, the embodied and the affective were modes of preserving memory that did not distinguish among the real, false, visible and invisible. We also conceived the project as a political strategy for the restoration of the silenced memory of a minority group and as a creative event in which the archive turned into a different performative act.

By working with the concept of *scriptive things*, we discerned in the archive multiple resonances between the disciplining of bodies and rhythms in the fac-tory and the disciplining of bodies and rhythms in higher education. Examples of this not only include the bell signaling the change of shifts in the factory, but also the student timetables that organize every single day at university. Addition-ally, when looking at documents found in the archive there are many resonances between the codes of conduct in the old factory and the current university. In our conversation with the women workers, we discovered that they often relaxed their obedience to the rules in moments of rest or when they shared snacks of bread and chocolate. This led us to ask how the students and I could transform our relationship to the university time-space, shaped by the linearity of time-tables evocative both of the factory and a neoliberal ideology of individualistic betterment and competition.

We decided to eat together in class while discussing preliminary ideas for the students' scores and the development of an archive of performances arising out of the Coma Cros inquiry.[4] This created a relational atmosphere that affectively and evocatively brought us closer to what Tamboukou (2016b) calls "moments of being" (85) when the workers collectively occupy their workplaces and populate them with their ideas, emotions, beliefs, habits and everyday practices, which in turn enables them to imagine these spaces differently.

The encounter with the former workers and our working lunch at the university opened up a third liminal space of understanding in which time present and time past collapsed (Dinshaw in Tamboukou 2016b, 80). The coexistence of different space-timings and urban rhythms influenced our debates and led us to ask what did it mean the everyday action of going to Coma Cros to work, study, teach and learn at the same place that had previously been inhabited by factory workers. It made us think about their echoes, ghosts and reverberations: An idea that strongly influenced the next stage of the project, in which the students envisioned, developed and enacted their own performances of the archive.

6.6 Embodied archives as contact zones

The visit to the municipal archive and the class in the street sparked a creative group process in which ideas, objects, stories and gestures were affectively collagedcollaged in a feminist practice the feminist practice of *speculative fabulation* (Blackman 2019, xii). These hybrid collages were used as performative strategies that allowed for the telling of more than one story at a time (Orr in Blackman 2019, 21). They brought together different fragments, including fiction, history, dreams, etc., in order to question linear time and disrupt patterns of connection and continuity. They activated imaginative restagings of gaps, silences and absences, both fictional and fantastic that unearthed fragments which captivated, attracted and inspired wonder in us. The topics, rumours and memories that we encountered in the archive functioned as vectors for possible paths of exploration. They directed us towards new configurations of what could be seen, said, thought, felt and imagined: A landscape of experiences taking shape between the past, present and future.

The students worked together to further develop their scores. First, they individually created their own collages of fragments and subsequently exchanged them with the whole group. Then they decided together the structure of the performances in an assembly-style process. Last, the pieces were completed in small groups. For this final phase, students explored Chow's (in Blackman 2019, 23) concept of *events of capture*. Events of capture are not judged for their truth-value or veracity but connected to experiences of *captivation* and processes of *contagion*. Thinking about *events of capture* spurred the students to consider their own investments in particular data. It brought them to consider potential contagions caused by their contact with the women workers. This gave rise to an understanding that different agents and materialities were entangled in the performance of the Coma Cros archive. For example, a presence of water and fluids in the landscape, but also in archive files and stories, sowed the seed for the writing of scores to proliferate in different directions and take on different themes. The women workers had shared stories of suicide and death in the river, the unbearable humidity of the factory and the lack of time for washing themselves and their undergarments during menstruation.

One of the performances, titled *Periods*, stands as another case of *scriptive things*. The performance recovers a lost repertoire concerning the ritual of cleaning menstrual blood. Figure 6.5 shows the traces of this performance, which

Figure 6.5 Image of the performance *Periods*. Courtesy of Assumpta Bassas.

could be seen as a ritual of restoration, cleaning the undergarments that the workers did not have time to wash. The thingness was coated with life by presenting the dignity of an intimate private object. *Periods* materializes the stained undergarments in the context of the oppression in Coma Cros and the silence of the archive; the underwear was a materialization of haunted data.

Elsewhere I have discussed the concept of *embodied archives as contact zones* (Vidiella 2014). This idea addresses accidental *becomings* as possible ways of documenting and creating an archive. The term *contact* is etymologically very close to *contagion* (both referring to touch). I explored the idea of archives of affects as spaces of circulation rather than repositories, that is, as shared flows where emotions act as force fields among bodies, subjectivities and objects, producing a distribution of intensities in continuous change. According to Braddock (2017), the concept of *contagion* does not reside in bodies or in things but rather operates as an atmosphere or *effluvia* (6). Through the concept of contagion, we can see data as phenomena rather than things or spaces. More specifically, Braddock affirms that understanding data as phenomena involves "performing the labor of tracing the entanglements, of making connections visible" (14).

Connected to this, in the first weeks of the unit we carried out a drift activity entitled *Looking for the Stage*, in which we walked silently around the neighbourhood seeking potential performance locations. As an example, many students chose balconies as their possible stages. This inspired a performance titled *Tagged*, in which a conversation took place between two balconies, with the two performers throwing the laundry hanging in the balconies into the street while talking about contagion among the human and non-human agents we had found in the archive files. These included the pollution and molecular particles that had caused diseases from pneumonia to thyroid cancer amongst workers in the textile factories. Also, the performers recited a list of the composition of the clothes, the city of manufacture and its shifts in textile manufacturing from cotton to polyester, and the places of production, from Salt to India and Pakistan.

During the drift, we interrogated banal domestic features such as shop windows and abandoned chairs and we became aware of sensory phenomena such as food smells, etc. The city's atmospheric archive helped us to approach Coma Cros as an ecology composed of the factory, the canal, the city, the Coma Cros neighbourhood, the people who had lived there before and were living there now, the traditions, the migratory diasporas and the gentrification taking place through the installation of the university. Kadja, an Afro-Catalan student, wrote in her portfolio:

> I want to highlight four things this drift made me think about. The first one was noticing the balconies and seeing the manifestation of the different social positions in Salt. Particularly, I could feel the political differences very strongly with the [hung] Catalan and Spanish flags. However, ... in the poorest ones there were no flags, they were just balconies either really empty or really full of stuff. The second thing that I sensed was how the art produced in Salt is hidden from the people who live here. I personally know some of the people living in this part of town. I know that their situation doesn't allow them to get ahead in some of the dreams or projects they have in mind. Perhaps because I knew this, I connected every graffiti or wall painting to someone's broken dreams.
>
> The third thing consisted in an involuntary event that happened when we crossed the street. We were walking in a straight line and in front of us there was a group of Black folks talking. I realized that my classmates unconsciously started to walk towards the other side of the street, crossing the road, and that while I was thinking about this I was following them. I was turning this event over in my mind for the rest of the drift. And lastly the dilemma we had amongst ourselves, creating a separation in the group, when some of us walked along one part of the street and a smaller group went in the other direction. I led a group that wanted to go towards a square in the centre of a park surrounded by blocks of flats, and in the end we all went that way. This event made us think about the relations of power in our group, reaching the conclusion that what we were doing involuntarily was to represent performatively what happened in this part of town every day.

This passage from Kadja's portfolio draws us closer to the idea of the *aliveness* of the archive and its data. It is an example of the city as living data and as a somatic archive. The artistic drift created an atmosphere of affective contagion that brought to the surface the social and power dynamics present in the group but also in the city.

6.7 Conclusions

In this chapter, I have discussed a pedagogical project to re-enliven and create archive assemblages relating to Coma Cros, an old cotton mill and currently the building where I teach. In doing this, I have sought to mobilize a number of concepts from the field of Performance Studies which I used to devise pedagogical situations and provocations for my students and to develop an affective inquiry into not only the Coma Cros archive but also the city and the specific place where they attend university. This process of inquiry highlighted the active role of materials in public life and their power to provoke effects and affects in participatory art-based archive research.

One of the things that most affected me in this pedagogical experience was how data became a contagious aesthetic potential; how objects, archives and bodies became possessed and haunted. This completely transformed my students' relationship with the pastness of the past and the multi-temporality of time. The aliveness of the data and the performativity of the archive propagated the sensation of non-linear timings in the city. The data had the *thing power* to affect the movements of bodies and their modes of inhabiting space, thereby generating the relocations and contagions evoked in the vignettes above. The *hauntological analysis* of the archive contributed to speculative endeavours to disrupt hegemonic historical continuities by focusing on art experiences that had the power to affect how we see and live with the archive, the city, the university and the factory as discontinuous and problematic entities.

There is an etymological connection between *haunting* and home, in the sense that haunting can propel feelings of not being at home in our surroundings. At the beginning of the project, the students felt disaffection with Coma Cros as a topic and place. They clearly saw that they did not want to be enrolled in a historical project as it felt dead, belonging to the past, unappealing and unfamiliar. This encouraged us to engage in a process of disorientation in Ahmed's (2006) sense of bringing different objects within reach, particularly those that might at first glance seem awry. The initial feeling of being out of place while occupying the same space changed gradually when we explored the archive with strategies like wonder in order to embrace odd, strange, anomalous and uncanny experiences. This created an *alien time* (Blackman 2019, 16) involving non-linear and entangled practices of memory and forgetting. In this sense, a pedagogy of the archive as an event may be a fertile terrain for arts-based researchers and artists who work with young people and groups emplaced in historical centres and who want to engage them in creative processes of inquiry into past–present relations on both community and personal levels.

Also, the current neoliberal situation of education and the tertiarization of urban spaces and buildings like Coma Cros, promoting creative economies and cultural industries, opens up an opportunity to redirect our attention to how performance and memory are taught, learned and explored in acts of in(ter)vention. By means of performance, the project conceived students' and workers' bodies as a threshold between the hegemonic narratives of capitalism and the power of imagination, understood as a "force that initiates something new in the process of archival understanding" (Tamboukou 2015, 156). At the same time, I am aware of the contradictions and traps that this experience involved. On the one hand, I orchestrated the educational conditions for students to attune to the material impact of gentrification and tertiarization, where industrial infrastructures have been turned into educational and cultural spaces that continue promoting precarious labour conditions. On the other hand, I am aware of being a gentrifying force (Denmead 2019), as I teach in a private college attached to a public university that prepares students (mostly white) to be part of creative economies in a non-white post-industrial city. While the tertiarized space of the university and its creative projects may revalorize the city, it is key to analyse how these projects continue making the city inaccessible to the majority of the second wave immigrants that populate Salt, as Kadja mentioned in her portfolio.

The *Pulling the Thread* performance series created by the students sought to address complex entanglements mattering together factory, creativity, aesthetics, the city, water-energy-force, women, education, work and feminist politics. It deterritorialized the class and the university. It opened up a field for lines of flight to be released and new subjectivities to emerge but with the tension in mind that these may soon be reterritorialized by the cultural industries or the academic system.

Notes

1 See the performances here: https://www.youtube.com/watch?app=desktop&v=czRE
   Wx0wV_4&feature=emb_logo
2 The *Confederación Nacional de Trabajo*, the National Workers' Confederation (CNT) is a Spanish anarcho-syndicalist labour union founded in Barcelona in 1910. It was strongly implanted in Catalonia, spurring workers to take over the means of production and especially to develop them; thus, it called for collectivization to promote the concentration of small factories and regulate production, successfully organizing many textile workers, most of them women.
3 The Art Factories or Cultural Factories project responds to the need for workspaces where artists can meet other creators. This is how former industrial buildings became venues for cultural and artistic activities in many cities in Spain. In 2007, to address artists' demands for suitably equipped premises, many Spanish city and town councils drew up the first official measures relating to the Art Factories. From that point onwards, they worked to build a network of publicly owned facilities.
4 According to Pablo Helguera,

> in the visual and performing arts the term *score* is borrowed from music to refer to a predetermined series of physical, verbal, or musical actions conceived by an artist and meant to be reinterpreted. The creation of a performance score in an artwork usually means that the work is not entirely ephemeral; the existence of a score means that it can be re-created or reinterpreted either by the artist himself or by a third party. In this sense the score becomes both a form of documentation and preservation of an artistic idea and a relatively flexible structure that usually allows a certain degree of interpretation of the work. (in http://intermsofperformance.site/keywords/score/pablo-helguera, consulted 7th June 2020)

References

Ahmed, Sara. 2006. *Queer phenomenology: Orientations, objects, others*. Durham and London: Duke University Press.

Barad, Karen. 2003. "Posthumanist performativity: Toward an understanding of how matter comes to matter". *Signs* 28, no. 3: 801–831.

Bennett, Jane. 2010. *Vibrant matter, a political ecology of things*. London: Duke University Press.

Bernstein, Robin. 2009. "Dances with things: Material culture and the performance of race". *Social Text* 27.4, no. 101: 67–94.

Bernstein, Robin. 2011. *Racial innocence: Performing American childhood from slavery to civil rights*. New York: New York University Press.

Blackman, Lisa. 2012. *Immaterial bodies. Affect, embodiment, mediation*. London: Sage.

Blackman, Lisa. 2019. *Haunted data. Affect, transmedia, weird science*. London and New York: Bloomsbury.

Braddock, Chris. 2017. *Animism in art and performance*. Switzerland: Palgrave Macmillan.

Chen, Mel. 2012. *Animacies: Biopolitic, racial mattering and queer affect*. Durham: Duke University Press.

Denmead, Tyler. 2019. *The creative underclass. Youth, race and the gentrifying city*. Durham and London: Duke University Press.

Harris, Anne and Holman, Stacy. 2019. *The queer life of things: Performance, affect, and the more-than-human*. London: Lexington Books.

Jones, Amelia. 2015 "Material traces: Performativity, artistic 'work', and new concepts of agency". *The Drama Review* 59, no. 4, winter: 18–35.

Larios, Shaday. 2018. *Los Objetos Vivos. Escenarios de la Materia Indócil*. México: Paso de Gato.

McKenzie, Jon. 2001. *Perform or else. From discipline to performance*. London and New York: Routledge.

Schneider, Rebecca. 2015. "New materialisms and performance studies". *TDR: The Drama Review* 59, no. 4 (T228), winter: 7–17.

Schweitzer, Marlis and Zerdy, Joanne. 2014. *Performing objects and theatrical things*, 118–131. Hampshire: Palgrave Macmillan.

Tamboukou, Maria. 2008. "Machinic assemblages: Women, art education and space". *Discourse: Studies in the Cultural Politics of Education* 29, no. 3: 359–375.

Tamboukou, Maria. 2014. "Archival research: Unravelling space/time/matter entanglements and fragments". *Qualitative Research* 14, no. 5: 617–633.

Tamboukou, Maria. 2015. "Feeling narrative in the archive: The question of serendipity". *Qualitative Research* 16, no. 2: 151–166.

Tamboukou, Maria. 2016a. *Gendering the memory of work: Women's workers' narratives*. London and New York: Routledge.

Tamboukou, Maria. 2016b. "Archival rhythms: Narrativity in the archive". In *The archive project: Archival research in the social sciences*, edited by Moore, Niamh, Salter, Andrea, Stanley, Liz and Tamboukou, Maria, 71–95. London and New York: Routledge.

Tamboukou, Maria. 2017. "Reassembling documents of life in the archive". *The European Journal of Life Writing* VI: 1–19.

Taylor, Diana. 2002. *The archive and the repertoire: Performing cultural memory in the Americas*. Durham and London: Duke University Press.

Ulmer, Jasmine. 2017. "Posthumanism as research methodology: Inquiry in the Anthropocene". *International Journal of Qualitative Studies in Education* 30, no. 9: 832–848.

Vidiella, Judit. 2014. "Archivos encarnados como zonas de contacto". *Efímera Revista* 5, no. 6: 16–23.

Part III
Ethics of Participation

7 The Lynden Sculpture Garden's Call and Response Program

To wonder, encounter and emplace through the radical Black imagination

Rina Little and Portia Cobb

7.1 An introduction

The processes, materials and embodiments by which place is enacted and performed shape experiences related to place. The stories we tell about these places also provide a framework by which to experience the world and craft conditions to know differently. This chapter tells a story that attends to the matter and materiality of the Lynden's Sculpture Garden's Call and Response Program (CRP), a museum programme enabling emplacement, experienced by one author, Rina Little, and imagined by the other author, Portia Cobb. "Call and response" is itself a format originating from many African traditions and present in the African diaspora, including that which was shaped by the slave trade. Often thought of as a pattern where one phrase is heard as commentary in response to another, here it is a programme that calls out to artists to co-create in response to artwork made, performances enacted and materials displayed at the Lynden. Through this programme and a variety of strategies, place is altered and connected to sites of struggle and to larger social, historical and political processes that (re)configure Black lives and presence in the city of Milwaukee.

Furthermore, the chapter elaborates on how as a process of creative inquiry the CRP does not separate researchers from research participants because we are part of the site's configuration. In our movement through and with the production and consumption of artworks, performances and material matters, we are part of and altered by the site. According to Barad (2007), the "apparatuses become the material conditions of possibility and impossibility of mattering" (148). Methods and outcomes inter-act. As Barad (2007) notes, researchers are

> not uncovering pre-existing facts about independently existing things as they exist frozen in time like little statues positioned in the world. Rather, we learn about phenomena – about specific material configurations of the world's becoming. The point is not simply to put the observer or knower back in the world (as if the world were a container and we needed merely to acknowledge our situatedness in it) but to understand and take account of the fact that we too are part of the world's differential becoming. (90–91)

Moreover, we write together as women of colour because we want to affirm a particular kind of politics in process, an alternative imagining. It is our job to do so because,

DOI: 10.4324/9781003027966-10

as Toni Morrison (1995) has said, it is "critical for any person who is black, or who belongs to any marginalized category, for historically, we were seldom invited to participate in the discourse even when we were its topic" (91). We write to enact an imaginary excised through relation and affect, as a "communal process of becoming" that includes witnessing, testifying, conversation and exchange (Sweet et al. 2020, 391). Since life exceeds representation, we must be attuned to a range of modalities that include affecting sensation and emotion to describe how we are involved and intertwined. These affective encounters are a crucial part of our knowledge production because they highlight the feelings of the other and for the other.

7.2 Racialization of spaces and a Black sense of place

As Lipsitz (2007) notes,

> The lived experience of race has a spatial dimension, and the lived experience of space has a racial dimension. People of different races in the United States are relegated to different physical locations by housing and lending discrimination, by school district boundaries, by policing practices, by zoning regulations, and by the design of transit systems. The racial demography of the places where people live, work, play, shop, and travel exposes them to a socially-shared system of exclusion and inclusion.
>
> (12)

In this chapter and through the use of art and social practice, we enact what a Black sense of place can be in the city of Milwaukee, Wisconsin, where the lived experience of race and space are connected. An analysis of US census data by the Brookings Institution, a nonprofit think tank, named Milwaukee the most segregated city in the nation (Spicuzza 2019). Black–White segregation is higher in Milwaukee than the national average, and affluent neighbourhoods sit next to the poorest neighbourhoods. In the midst of such an environment, the Lynden Sculpture Garden created a programme in 2016 called the CRP. The CRP forges a Black sense of place with artists in a segregated city.

As McKittrick (2006) contends, approaching space and place as "merely containers for human complexities and social relations is terribly seductive" for it "seemingly calibrates and normalizes where and therefore who, we are" (xi), displacing difference and erasing bodies. But we do not think spaces and places are containers; they are produced through social and material practices that organize, build and imagine our surroundings. We were drawn to the CRP because it also conceptualizes place as a location through which movements, interactions, materials, communications and affects create knowing. By learning in and through art as material matter in the CRP, place is produced specifically to challenge traditional ways of knowing, to render places and histories tangible and to provide spaces where cultures can meet and exchange multiple and contested stories. The programme offers a way of being political through the affective.

Being political through the affective envisions and mobilizes micropolitical acts that enhance and intensify the relational capacities among bodies, places, stories and things. In the CRP, these things included buildings, objects, foods and indigo ink to catalyze a new imagination of place where Black spatial relations and embodiments are at its centre. As Bennett (2010) has noted, such a vision of affective politics involves not only paying attention to hard political principles and moral mandates that "risk just being a bunch of words" (xii) but underexplored ethical sensibilities and affective atmospheres, and how they are also agentive in (re)forming social relations. The identities of Blacks in the US and the Caribbean are dramatically grounded to the land through labour but also through song, art and the imagination, a relation that has continued over time as a nurturance of legacy and cultivation. However, as Finney (2014) has argued, the ways land and the experience of environment has been conceptualized in the US as national discourse is informed by White European values around conservation and revitalization that reflects little on issues of access, privilege and other material and environmental experiences of Black Americans and their connections to land. This means that building a Black sense of place involves an engagement with a recognition of the erasure created by current spatial arrangements and creative practices involving Black modes of place-making and place-imagining, "open(ing) up the possibilities of thinking collectively about the production of space as unfinished" (McKittrick 2006, xiii).

Building on these ideas, we articulate our understanding of arts-based inquiry in connection to histories and aesthetic practices connected to the Black diaspora, and we reflect on them as a series of material, aesthetic and social acts of place-making constructed through the CRP, of which we are a part. Rina is an educator in relation to the programme, and Portia is an artist in the programme. The Lynden Sculpture Garden is situated in the north of Milwaukee, and it houses a collection of sculptures, trees and vegetation on the grounds of what was once the estate of industrialist Harry Bradley and his wife Margaret Bradley. The Bradley home opened to the public in 2010 as a museum and staff began to work towards place-conscious programming under the leadership of Polly Morris. The central goals of the space are to act as laboratory and resource for artists, to forge dialogues and partnerships between artists and communities, to integrate sculpture with the ecology of the land and to reframe the permanent collection through contemporary work and issues. Institutionally, the programming of the CRP has significantly engaged in a decolonization of Lynden as a place and as land. Initially attached to White philanthropist values of conservation and beautification by the Bradleys, who collected both trees and sculptures from all over the world and placed them on their private estate, the CRP has resonated with the land differently. It permits experiences with the space through material and social practices that are speculatively in tune with a Black presence not recognized before (Nxumalo 2018). This is an important political gesture in a city like Milwaukee where Blackness and space can be easily connected to dispossession.

The CRP is a space and practice that gathers together a community of artists working across disciplines who are committed to the radical Black imagination. Black radical thought seeks to claim and reimagine the city by creating alternative social and spatial transformations through progressive action to meet the needs of Black people (Tyner 2006). This is a space of wonder, of cultivation, of healing, of collective freedom, where the imagination is understood as something that can transform the world and produce conditions that support it. The programme is a means to re-examine the past and propose a future that actively addresses citizenship and belonging using form and content (Lynden Sculpture Garden n.d.-a, "Call and Response"). It fits the Lynden's goals of acting as a laboratory and resource, forging dialogue with the community and reframing its collection with contemporary work and issues.

7.3 Utilizing Black geographies

As Nxumalo (2016) proposes, it is important to critically consider the worldviews and ontologies that we use to think, approach and involve others in research about place. With this in mind, in this chapter, we engage with the field of Black geographies because the field proposes interrelations between geography, Black studies and the arts, as ways of generating "contestations of place in view" (8).

Traditional geographies believe that space is transparent and knowable in geometric terms, but some scholars have challenged that. Massey (2005) suggests that space should be understood as contingent, active and under construction through social and material relations and practices. According to her, places are not things but events involving humans, nonhumans, materials and matters that are thrown together. Ingold (2000) goes on to propose that a place is not surroundings but a zone of entanglement produced from movements with varying intensities. Space is produced through interactions.

Many believe that traditional geographies cannot do emancipatory work because those involved embrace positivist philosophy and universal spatial laws, marginalizing, dismissing and erasing others and their epistemologies through displacement and ignoring the material realities of differential embodiments. McKittrick (2006) states:

> while the power of transparent space works to hierarchically position individuals, communities, regions, and nations, it is also contestable—the subject interprets, and ruptures, the knowability of our surroundings. What this contestation makes possible are "black geographies," which I want to identify as "the terrain of political struggle itself," or where the imperative of a perspective of struggle takes place.
>
> (6)

She goes on to explain how the discourse of transparent space has also excluded Blackness or has subordinated this knowledge and experience in a deliberate attempt to destroy a Black sense of place.

In examining the production of space in the CRP, we utilize Black geographies and Black ontologies, where race, gender, sexuality and class are understood to determine inhabitation and re-inhabitation (McKittrick 2011). Black geographies emerge partly from Black radical thought (Moten 2003) and use Blackness and race to critically contribute to geography. Black radicalism refers to militant politics and thought that challenge exploitation, the archive, structural and material dispossession, social inequality, marginalization, and private and state-sanctioned anti-Blackness and rethinks histories and afterlives of Middle Passage, racial capitalism and settler colonialism. Black geographies expose the limitations of understanding space as transparent and tap into the material, the imaginary, the philosophical and the representational to produce alternative patterns of knowing (McKittrick 2006). Black geographies acknowledge that social identities of bodies operate spatially, providing insight into how people have had to negotiate places, spaces and themselves (Gilmore 2005; Pulido 2006; Schein 2006; Lipsitz 2011; Shabazz 2015).

Furthermore, discussions of the ontologies of Blackness have reshaped our consideration of what it means to be human, exposing its fragility as a concept. According to Fredrickson (2002), race has always distorted the concept of human separating some from others. This difference provides a motive for treating some as less than human, or nonhuman. Moreover, spatial practices within racism keep Black cultures in place often fixing them to particular spaces and times and defining them as placeless and nonhuman (McKittrick 2011). In doing so, Eaves (2017) warns, "the study of racialized others and their habitats, social relationships, and economic contributions become merely sites of containment, rather than sources of important geographic information" (84).

As a source of geographic information, Black geographies also offer alternative tellings and different spatial imaginaries of the world expanding how space can be produced (McKittrick and Woods 2007; Eaves 2017). Black people who have been abducted through transatlantic slavery and made part of the plantation economy have always contributed to the production of space not only through their labour, but also through song, storytelling and the imagination (Finney 2014). McKittrick (2011) discusses how various practices through history erased a Black sense of place so that Black bodies could be targeted as part of a plantation economy, which included the slave plantation, the auction block, the big house, the fields and crops and the slave quarters. This plantation economy dehumanized Blacks and normalized Black dispossession and White supremacy. But such bondage did not prevent enslaved people from defining the land and writing their world. Fugitive and maroon maps, food-nourishment maps, family maps and music maps were all produced, along with side road, route and boundary maps. Furthermore, Blacks invested in the making of place through alternative material and imaginary spaces. Toni Morrison (1995) constitutes a paradigmatic example. She uses literary archaeology to journey to a site to see what remains there and to reconstruct the emotional memories that these remains imply. Saidiya Hartman (2008) also uses such critical fabrication, but in her case, she combines archival research with a fictional narrative to make the absent voices of enslaved women heard through storying. Both writers create a space of resistance, imagined for change. In addition, there are also methodological practices in place-based research that Nxumalo (2018) calls "testifying-witnessing" (14). Borrowing from Black feminist practices (Tarpley 1995; Collins 1998), Nxumalo (2018) says testifying-witnessing makes "visible the complexities of Black geographies beyond stories of damaged place relations, surveillance and absence" (14). Testifying-witnessing is a relational practice that includes affective responses situated within particular Black experiences and actively names multiple truths, including injustices and affirmations of strength and resilience, using creative means.

We frame our own engagement with the CRP within the Black methodologies of place proposed by Black artists and scholars Morrison, Hartman, Nxumalo and others. Thus, we consider enacting, experimenting and inventing with such methodologies can lead to possible reconfigurations in the city that affirm Black presence. The CRP reappropriates space as a social and material transformation in order to claim and to reimagine the city complexly. The land is not understood as a flat terrain but as a deep space (Massey 2005). The CRP constructs this as a space in which to dwell on a multitemporal scale where Black presence rooted in the past, present and future can be felt. Furthermore, participants in the CRP increase the use value of the space through congregation and inhabitation. According to Trigg (2012), "inhabiting" refers to the interplay between ourselves and the places we find ourselves. The result of this shared inhabiting is the hybridity of place and self, which allows us to see differently. As McKittrick (2011) relates, "black sense of place can be understood as the process of materially and imaginatively situating historical and contemporary struggle against practices of domination and the difficult entanglements of racial encounter" (949). A Black sense of place is thus not the authentication of Blackness nor an offer of a better place for Blacks but instead refers to a space of encounter that holds inside itself useful anti-colonial practices, narratives and resistance. The CRP examines alternative forms of living through undoing and rearrangement, including that of materials (Thompson 2017).

7.4 The Call and Response Program as a different space of encounter with nature

The Lynden features a type of landscape garden that emerged among British noble and upper middle classes in the early eighteenth century, people that financed the industrial revolution by investing in the slave trade and forced labour in overseas plantations. This model of the garden was also adopted by industrialists in America, like the Bradleys. Connecting themselves to this heritage, the Bradleys constructed the Lynden as an idealized view of pastoral nature with ponds, bridges and rolling lawns set against groves of trees imported from all over the world and replanted at the estate. In the garden, the city disappears and art emerges. The permanence, the noise, the population density and the fast rhythms that make up the city move to impermanence as trees and vegetation change. The white noise of cars is replaced by an idyllic quietness only interrupted by birds tweeting, frogs croaking or leaves moving. Few people linger on the art in a relatively still and empty outdoor space. The natural scenery resembles that found in landscape paintings, blurring the boundaries between art and nature with the picturesque, a spatial imagery where Blackness is rarely present.

As one moves towards a work of art entitled *Eliza's Peculiar Cabinet of Curiosities* (2016), a slave cabin which initiated the creation of the CRP, one crosses this scene, where the collection of monumental sculptures can also be found and travels across the garden to arrive into a wooden thicket located at an outer limit, where artefacts made and performances staged during the year for the CRP can be experienced. Situating the art of the CRP at the outer limit of the English Garden is an important choice. It makes explicit that the garden is an enclosure and evokes the understanding that slave cabins were always located in marginal spaces of the plantation away from its vistas. The English Garden and many of the artworks made through the CRP aims to connect people to the natural world. However, as Finney (2014) has noted, the history of conservationism in the USA even that connected to national parks has normalized notions of environmentalism that excludes that exercised by Blacks. Conservationism builds itself on idealistic notions of nature where the land is separated from histories of conquest, colonialism and proprietary regimes. In turn, this has led to ignoring the messy and imperfect nature systems that are already present in urban spaces and limits understandings around what counts as nature and life in the city (Nxumalo and Ross 2019). According to McLean (2013),

> white bodies become white through the essentialization of nature as a pure uninhabited space. In this way, nature is constructed as a cleansing system, a place where white bodies can escape the negative consequences of urban industrialism, and reclaim identities of innocence…These discourses work to produce environmentalism as a space where white identities safeguard and maintain the land, rather than consume and destroy it.
>
> (360)

Furthermore, environmental organizations have often framed the absence of Blacks from their programmes as rooted in a lack of interest in nature, but this is a continuation of the colonial perspective that regards White people as the custodians of nature. Various reasons have kept some away from nature including associating it with physical work, poverty, legacies of slavery, colonialism and limited options (McLean 2013).

The CRP reconnects visitors both young and old to nature and the land, allowing them to enjoy the aesthetics of the outdoors and uses art to feel (dis) connected with natural and cultural systems. It does not position nature as a space for play and discovery for some and as a form of rescue for individual development, academic outcomes in science and a respite from rough neighbourhoods for others (Nxumalo and Ross 2019). Although experiential, the CRP also connects to social and political issues without essentializing nature and reinforcing binary views of nature. In reimagining Black space at the Lynden as part of an arts-based environmental education, participants delve into speculative fiction as an imaginative emancipatory movement that disrupts the dislocation of Blacks from natural spaces. It is counter to connecting Whites to pure nature. Instead, it offers a way of seeing and imagining the various ways in which Black lives are lived. It imagines Black futurities in response to settler colonialism, anti-Blackness and Black erasure by re-envisioning the relationship between Blackness and natural spaces. As Nxumalo and Ross (2019) remind us, play, embodied encounters with the outdoors, humour, activism, environmental science, discussion on environmental racism, history and geographies are all necessary to Black spaces.

7.5 The Call and Response Program constructs place through processes and experiences

The CRP is a unique space to study how to enable investment in and security of a place. Here, contemporary art is used to facilitate this through material encounters. The institution invites artists to make art that can become a catalyst for new, entangled relations among people, communities and places (Peers and Brown 2003). We, as an educator and an artist, have accompanied visitors as they wander across the lawn of the Lynden, pass modernist sculptures, experience edge effects, cross a bridge over an artificial pond with frogs croaking and enter a wooded area. Tucked inside a clearing is the slave cabin, with a vegetable garden and bits of prairie grass (Figure 7.1). Visitors enter the cabin, which has an open front door and side wall. All are welcomed inside by this spatial manipulation and encounter many objects. As they dwell within the space, they come to know aspects of the occupant's life through her objects—a field journal with entries and drawings; a small writing desk; a ceiling lined with facsimiles of the *Emancipation Proclamation*; imagery from Aesop's *Fables*; photographs; sculptures; specimens; books; bones; jars; video of rippling water; and more. The cabin resonates with the life of an intellectual from the antebellum period. But there are objects that puzzle and do not seem to fit the time period, such as a Princess Leia action figure or a sculpture made of arms and legs of a variety of decaying dolls. These bits and pieces seem out of place and the clarity of the scene in which they obtrude is impaired by their presence, making us pause. *Eliza's Peculiar Cabinet of Curiosities* (2016) is the work of contemporary artist Folayemi Wilson. Wilson's character of Eliza is an African-American slave, a scientist, a time traveller, an artist, a collector and a writer; she lives "here" and travels "there" making a study of the world around her. The house demands attention since it is built to

Figure 7.1 Eliza's Peculiar Cabinet of Curiosities (2016). Image courtesy of Jim Wildeman.

reimagine, offering an in-betweenness of spaces of past-present-future for the intersection of multiple stories that contest what we imagine of Black women, of slave women, of time-space matters.

To highlight a commitment to a place that is culturally and environmentally responsive, the Director of the Lynden, Polly Morris, invited Wilson to create an installation as a full-scale Wunderkammer, literally a "room of wonder" or cabinet of curiosities, and slave cabin. As McKittrick (2006) asserts, spatial manipulation using materials makes possible interrelated processes that produce a different type of picturing of Black women. Instead of servitude, inhuman being, worker, captive, or an objectified site of sex, violence and reproduction, the work imagines what the empowered nineteenth-century woman of African descent might collect and includes objects from captors, plantation life and the natural world. It is what Nxumalo (2018) calls a testifying-witnessing where the means by which to imagine the invisible is made visible through speculative fabulation, storying as a way of crafting relations of possibility (Haraway 2016). Eliza explores her own voice through her field journal writings and drawings, collections and displays when historically people like her were rendered silent and this affects us intensely through her material processions by having us consider what was and what could be possible.

The fictional Eliza makes material her world using creative forms of social commentary. She is in touch with her surroundings past, present and future, including the natural space. Here, nature is positioned as a space of discovery and scientific exploration, a space often denied to Blacks and others from historically marginalized communities. As Nxumalo and Ross (2019) remind us, imagining the production of Black spaces necessitates a re-envisioning of the relationship between Blackness and nature spaces. This Black woman is visually and socially represented in the landscape and is made a viable contributor to an ongoing geographic struggle for space and place. The architecture, material objects and natural specimens become important agents for the inclusion of voices that are usually marginalized (Lynden Sculpture Garden n.d.-b, "Fo Wilson: Eliza's Peculiar Cabinet of Curiosities"). The invention of Eliza creates an "alterable terrain in which Black women assert their sense of place" (McKittrick 2006, xviii).

This cabin also anchors the CRP as a site where artists from multiple disciplines develop inquiry and social practice while sharing a commitment to the radical Black imagination to gather together, invent and create a sense of belonging. What began as a unique work of art of Black presence on the Lynden grounds turned into a way of programming by the Fall of 2016. Wilson first called to spoken word performers to respond to her cabin and resulting programmes became cross-disciplinary, focused on community and centred on voicing artists of colour. According to Lynden Director Morris, the programme "required us to create and sustain relationships among artists, to look beyond Lynden's borders, and to stitch ourselves into our community" (2017, n.p.). Later, Wilson would also call on visual artists to respond to her work and they in turn called on others. It is currently in its sixth year of programming and includes performances, artworks and installations, workshops and symposia.

As an Afrofuturist artist, Wilson decolonizes and re-inhabits by offering an alternative history worth recovering. Afrofuturist artists use a range of media and draw on the speculative or speculative fabulation to examine, critique and revise historical understandings of lives and events, and to (re)imagine the future (Derby 1994; Haraway 2016). Wilson uses the imagination to design alternatives to transform lives. Common characteristics of Afrofuturism include: (1) going forward by first going backward into the past; (2) presenting counter memories; (3) exploring magic realism; (4) combining elements of science fiction, fantasy, history, Afrocentricity and/or nonwestern cosmologies; (5) envisioning, shaping, managing and delivering counter-narratives (the gaze); (6) using imagination to transcend circumstances and empower; (7) closing the digital/tech divide; (8) providing a space for Black women to engage with the intersections of race and gender; and (9) seeking to disturb or defamiliarize through creative processes (Derby 1994; Yaszek 2006; Rambsy 2012). The Afrofuturist artist focuses on how Afrofuturism can be a lens through which to investigate belonging, or to investigate being set in place by presenting a counter-narrative through, for example, magic realism, science fiction, fantasy or history, which speaks to the call for alternative tellings that entangle facts and fictions and imaginatively situate historical and contemporary struggles. The site is both material and discursive, presenting psychological, mythic and historical connections. Wilson shares:

> The project foregrounds Eliza's experience—fictional or not—and the imaginations of others like her as a unique technology of Black agency, resistance, and survival, and as the under-appreciated gift of Blackness from which all of America has benefitted.
>
> (Wilson 2017, n.p.)

Wilson's cabin curates the life of Eliza. It is a microcosm of a world and of memory, mixing fact with fiction, showcasing the inhabitant's education. It takes visitors back into the past so that they can reimagine a future from the perspective of the Other. Eliza's cabin and its surroundings vibrate with sensorial experiences that counter the normative knowledge of enslaved, Black women. Viewers instead witness a becoming, where the materiality of the environment gives Eliza memories and histories and in turn acts upon us and upon others who encounter it. If people, animals, plants and objects act, react and become with one another, we are also required to be obligated to the other and to move to co-creative relationships with these presences (Greeson 2016; Caniglia 2018).

7.6 Relational movements in response to the multisensory and juxtaposed environment

Author, filmmaker and installation artist Portia Cobb's *Rooted: The Storied Land, Memory, and Belonging* responds to Wilson's call to connect herself to the cabin. *Rooted* was a multi-year artist residency which involved making a garden, performances with the community and collaborations with others. Cobb recalls:

> My work memorializes food ways that survived slavery through the use of a vegetable garden cared for by Lizzie, a woman born free, with a relationship to Eliza. She is based on my Great Aunt Lizzie. The garden is a reminder of the planation as a site of production from which a racial capitalist system

was built but also a space where "survival, substance, resistance and affirmation" are born (Woods 1998, 27). The stories I tell and share of home and place have evolved over my lived experiences with movement set in motion when I was a small child. My mother left her birthplace, Charleston, South Carolina at the age of 21. She lived in Boston for a short time and then made her way to New York where she had familial connections. She worked and sent money back home "down south" to help with family and to support the new home her father was building that became our homestead.

My mother would remind those she met that we were from the East sometimes calling out New York but at other times, South Carolina was named. South Carolina holds a place for me because through my mother's bloodline, I belong to a cultural and ethnic diaspora of Gullah Geechee people, descendants of Africans brought to the Carolinas. My creole identity is solidified through linguistic, culinary, and other cultural continuums. We ate rice, seafood, okra, field peas, boiled green peanuts, and an assortment of greens, which reinforced the memory of our Gullah Geechee identity. My mother took pride in our past and those surroundings that I witnessed manifest in the stories I tell as an artist and the representations I create of home and place. I layer my history through the imaginary as discussed in black geographies where the site of memory is also the sight of memory (McKittrick 2006).

Instead of a plantation which violently tied Blacks to the land and the historical trauma of it, Cobb presents a garden that re-envisions production as creative and nourishing. The garden is Lizzie's but also called the Emancipation Subsistence Garden (Figure 7.2). Cobb invites others to respond to both Lizzie and Eliza, including the anthropologist, food scholar and chef Scott Barton, who pickled produce in 2018 from the garden and organized a harvest story table (Figure 7.3). Barton became interested in southern food and the contribution of Africans and African Americans to the culture of food and foodways while living in New York. He worked with Cobb and other Call and Response artists on the harvest story table which explored diaspora legacies through food and foodways and engaged a group of elders in discussions about land, gardening, migration and food. Her invitation to respond to her garden also includes textile artist Arianne King Comer, who worked with the community to make a cloth recipe book of foods from the garden using indigo inks and dying techniques (Figures 7.4 and 7.5). Furthermore, a dinner was hosted by the Lynden and included 45 invited guests and such eighteenth-century menu and re-envisioned items, as shrimp paste benne wafers, okra and watermelon rind pickles, red rice, fritters, boiled peanuts, spoonbread, hominy, field peas, collard greens with kimchi, crab rice, rice pudding, sweet potato pone and sweet and sour peaches. Cobb and Comer King researched recipes using oral histories, archives and family folklore.

All these entanglements are co-creations, calling us into connection with a web of movements, actions and materials that construct modes of emplacement. It is what bell hooks (1995) called an aesthetics of existence, a performative act remaking existence; it is a way of inhabiting space, a particular location; it is a

Figure 7.2 Portia Cobb in "Lizzie's Garden". Image courtesy of Maeve Jackson.

Figure 7.3 Scott Barton with guests at harvest table dinner in 2018. Image courtesy of Lynden Sculpture Garden.

Figure 7.4 Arianne King Comer leads a workshop in 2018. Image courtesy of Sara Risley.

way of looking and becoming through oppositional politics where we are called to witness, to listen and to meet the other on their terms. As Cobb explains,

> My practice is to uncover, collect, contextualize, and re-imagine. *Rooted* was also a project that allowed play with elements in an open space. In 2018, I gathered together 40 visitors between the garden and the cabin, sharing the space with the vegetables [Figure 7.6]. I turned the space into a portal to the past by transporting visitors to Yonges Island, one of the Sea Islands and a chosen home for former slaves. My ancestors settled there. I also spent child-hood summers there. As I sat on the platform of the cabin and watched the dancers, I embodied Lizzie and voiced how the food from the growing garden sustained a growing community during the Reconstruction era.

The objects, images, sounds, smells, tastes, touch and the experiences produced are in movement and in relation to each other, "generated through their inter-relatedness with both the persons they move with and the environments they move through and are part of" (Pink 2011, 4). These acts intensify relations among bodies, stories and things. They emerge from and are implicated in the production of place as matters "in-motion, in-relation" (Nxumalo 2016, 41).

Figure 7.5 Community participants at Arianne King Comer's workshop. Image courtesy of Sara Risley.

Figure 7.6 A performance connected to *Rooted: The Storied Land, Memory, and Belonging* at *Eliza's Peculiar Cabinet of Curiosities* in 2018. The Jazzy Jewels, a troupe of elders, who perform a line dance, join Portia Cobb who brings Lizzie to life. Dresses are designed by Arianne King Comer. Image courtesy of Sara Risley.

7.7 Eliza's Peculiar Cabinet of Curiosities reconsidered through wonder

As an organizational structure, the cabin, as a cabinet of curiosities or wonders, provides a generative, multilayered and comprehensively significant interactive framework for emphasizing the materiality of the process and labour of emplacement (Huber and McRae 2019) and is part of a relational field of interwoven entanglements (Ingold 2000). Historically, cabinets of curiosities were a means of display for the wealthy and powerful as they curated their collections; these were then later linked to museum displays. However, such cabinets differ from modern museums in their emphasis on the importance of the curator. Lubar (2017) notes:

> The earliest museums, the early modern cabinets of curiosities, were, like museums today, collections of artifacts on shelves and in cabinets. But cabinets of curiosities had a different point to make. Early cabinets, in the sixteenth and seventeenth centuries, were heterogeneous, unsystematic displays. The message they sent was one of exoticism and variety. Look at all of these amazing things, they said. What a worldly person the collector must be! Each object stood for itself.
>
> (179)

Eliza is not wealthy, of course, but her collection is unique, specific to her, and thus emphasizes her importance in its curation. Furthermore, as a catalyst to the CRP, it emphasizes a multi-vocal process that troubles hierarchical representations of knowledge. Its display does not isolate objects for contemplation but promotes unusual juxtapositions in which a multiplicity of possible meanings can emerge through relation, activity and calls for response. Invited artists and audiences are asked to interact with it and through multiple points of connection. The Wunderkammer is a wonder cabinet. As Greenblatt (1990) explains,

> [the] wonder-cabinets of the Renaissance were at least as much about possession as display. The wonder derived not only from what could be seen but also from the sense that the shelves and cases were filled with unseen wonders, all the prestigious property of the collector. In this sense, the cult of wonder originated in close conjunction with a certain type of resonance, a resonance bound up with the evocation not of an absent culture but of the great man's superfluity of rare and precious things.
>
> (29)

The CRP uses wonder to expand the radical Black imagination. MacLure (2013) discusses wonder as both material and relational. It exists in body and mind, "emanating from a particular object, image, or fragment of text", crafting the challenge, "what next?" (MacLure 2013, 229). Furthermore, wonder as material provides a way to highlight interconnected relations and a sense of place. It is not only a place where objects are housed but also a setting where relationships are made. The CRP provides an environment to grapple with the complexities of how art maintains, transforms and creates ways of understanding in relation to people and facilitate place-conscious inquiry and relation to the city through a recombinatory process using materialities.

The CRP gathers together artists as a means to re-examine the past and imagine a better future and to address belonging through decolonization and re-inhabitation. It is a space where artists investigate complex issues, collaborate and extend dialogue, and their productions become catalysts for new relationships and knowledge constructs. The Lynden's CRP also facilitates interactions with participants. Perspectives of the museum, artists and participants intersect with and inform each other. In this zone, people have reason to come into contact with each other because they have aims that extend beyond the borders of their disciplines, to imagine new public spheres and relationships with place through the radical Black and as a site of struggle, testimony, witnessing and work. The CRP stretches beyond preserving and contextualizing objects and instead acts as a resource of materiality for rewriting histories and social relationships with surrounding communities. Contemporary art practices at the CRP are used to inquire how spaces could be re-inhabited differently. People and objects become a function of the space. Black geographies allow for spatial understandings that revitalize space through the imaginary. In the CRP, participants are included in social interactions surrounding the production and reception of place through artist practices in order to connect people and ideas to their surroundings and to understand creative modes of emplacement, exchange, negotiation and communication (Gere 1997).

There are a number of questions that remain open that we continue to think about as we dwell with the CRP methodology of call and response. How can the wondrous force of contemporary art practices as radical forms of Black inquiry and socialization avoid reification, appropriation or tokenization when inhabiting museums? How can invitations to non-Black Milwaukeeans to dwell in these spaces and relations provoke further engagement in troubling the ongoing settler-colonialist practices with which they are intimate in their daily life in the city? How do emplaced methods based on lived experiences of encounter, wonder and inhabitation remain open and not become knowledge that speaks for Blackness or constructs erasure? Inspired by Nxumalo (2018), what we think is that the CRP as an ongoing project and methodology needs to continue framing and reframing the tensions embedded in such questions by enacting space in deep and contested ways (McKittrick 2006). This requires persistence in a continued process of building place experiences that are intricate, complex, multiple, temporal and defined by frictions since the CRP also exists within a neoliberal system. It seems that the ongoing nature of the CRP programme holds great potential to engage and re-enact Black methods of connection with the land, posing resistance to possession, control and knowability while enacting and experimenting with new potentials for feeling, sensing, living Black presences of the past, present and future.

References

Barad, Karen. 2007. *Meeting the universe halfway: Quantum physics and the entanglement of matter and meaning*. Durham: Duke University Press.

Bennett, Jane. 2010. *Vibrant matter: A political ecology of things*. Durham: Duke University Press.

Caniglia, Noel. 2018. "Outdoor education entanglements: A Crone's epiphany." In *The Palgrave international handbook of women and outdoor learning*, edited by Tonia Gray and Denise Mitten, 461–471. New York: Palgrave Studies in Gender and Education.

Collins, Patricia Hill. 1998. *Fighting words: Black women and the search for justice*. Minneapolis: University of Minnesota.

Derby, Mark. 1994. "Black to the future: Interviews with Samuel R. Delany, Greg Tate, and Tricia Rose." In *Flame wars: The discourse of cyberculture*, edited by Mark Derby, 179–222. Durham: Duke University Press.

Eaves, Latoya. 2017. "Black geographic possibilities: On a queer black south." *Southeastern Geographer* 57, no. 1: 80–95.

Finney, Carolyn. 2014. *Black faces, white spaces: Reimagining the relationship of African Americans to the great outdoors*. Chapel Hill: The University of North Carolina Press.

Fredrickson, George. 2002. *Racism: A short history*. Princeton: Princeton University Press.

Gere, Charles. 1997. "Museums, contact zones and the internet." In *Museum interactive multimedia 1997: Cultural heritage systems design and interfaces*, edited by David Bearman and Jennifer Trant, 59–66. Paris: Archives & Museum Informatics.

Gilmore, Ruth. 2005. *Golden gulag: Prisons, surplus, crisis, and opposition in globalizing California*. Berkeley: University of California Press.

Greenblatt, Steven. 1990. "Resonance and wonder." *Bulletin of the American Academy of Arts And Sciences* 43, no. 4: 11–34.

Greeson, Kimberley. 2016. *Pollinators, people, plants: A multispecies ethnography of the biopolitical culture of pollinators in Hawaii*. Unpublished manuscript. Sustainability Education Doctoral Program, Prescott College, Prescott, AZ.

Haraway, Donna. 2016. *Staying with the trouble. Making kin in the Chthulucene*. Durham: Duke University.

Hartman, Saidiya. 2008. "Venus in two acts." *Small Axe Journal* 12, no. 2: 1–14.

Hooks, Bell. 1995. "An aesthetics of blackness: Strange and oppositional." *Lenox Avenue: A Journal of Interarts Inquiry* 1: 65–72.

Huber, Aubrey and Chris McRae. 2019. "Wunderkammer: The performance showcase as critical performative pedagogy." *Text and Performance Quarterly* 39, no. 3: 285–304.

Ingold, Tim. 2000. *The perception of the environment*. London: Routledge.

Lipsitz, George. 2007. "The racialization of space and the spatialization of race: Theorizing the hidden architecture of landscape." *Landscape* Journal 26, no. 1: 10–23.

Lipsitz, George. 2011. *How racism takes place*. Philadelphia: Temples University Press.

Lubar, Steven. 2017. *Inside the lost museum: curating, past and present*. Cambridge: Harvard University Press.

Lynden Sculpture Garden. n.d.-a "Call and response." August 1, 2020. https://www.lyndensculpturegarden.org/callandresponse.

Lynden Sculpture Garden. n.d.-b "Fo Wilson: Eliza's peculiar cabinet of curiosities." August 2, 2020, https://www.lyndensculpturegarden.org/exhibitions/fo-wilson-elizas-peculiar-cabinet-curiosities

MacLure, Maggie. 2013. "The wonder of data." *Cultural Studies Critical Methodologies* 13, no. 4: 228–232.

Massey, Doreen. 2005. *For space*. London: Sage.

McKittrick, Katherine. 2011. "On plantations, prisons, and a black sense of place." *Social & Cultural Geography* 12, no. 8: 947–963.

McKittrick, Katherine and Clyde Woods. 2007. "No one knows the mysteries at the bottom of the ocean." In *Black geographies and the politics of place*, edited by Katherine McKittrick and Clyde Woods, 1–13. Toronto: Between the Lines.

McKittrick, Kathrine. 2006. *Demonic grounds: Black women and the cartographies of struggle.* Minneapolis: University of Minnesota Press.

McLean, Sheelah. 2013. "The whiteness of green: Racialization and environmental education." *The Canadian Geographer* 57, no. 3: 354–362.

Morris, Polly. 2017. *Director's note: Call and response.* Milwaukee: Bradley Family Foundation, Inc.

Morrison, Toni. 1995. "The site of memory." In *Inventing the truth: The art and craft of memoir*, edited by W. Zinsser, 83–102. New York: Houghton Mifflin.

Moten, Fred. 2003. *In the break: The aesthetics of the black radical tradition.* Minneapolis: University of Minnesota Press.

Nxumalo, Fikile, 2016. "Storying practices of witnessing: Refiguring quality in everyday pedagogical encounters." *Contemporary Issues in Early Childhood* 17, no. 1: 39–53.

Nxumalo, Fikile. 2018. "Situating indigenous and black childhoods in the anthropocene." In *Research handbook on childhoodnature*, edited by Amy Cutter-Mackenzie, Karen Malone and Elisabeth Barratt Hacking, 2–19. Switzerland: Springer, Cham.

Nxumalo, Fikile and Miraya Kihana Ross. 2019. "Envisioning black space in environmental education for young children." *Race, Ethnicity and Education* 22, no. 4: 502–524.

Peers, Laura and Alison Brown. 2003. Introduction. In *Museums and source communities*, edited by Laura Peers and Alison Brown, 3–6. London: Routledge.

Pink, Sarah. 2011. "Sensory Digital Photography: Rethinking 'Moving' and the Image." *Visual Studies* 26, no. 1: 4–13.

Pulido, Laura. 2006. *Black, brown, yellow, and left: Radical activism in Los Angeles.* Berkeley: University of California Press.

Rambsy, H. April 2012. "A notebook on afrofuturism." *Cultural Front.* December 1, 2020. http://www.culturalfront.org/2012/04/notebook-on-afrofuturism.html

Schein, Richard. (2006). Race and landscape in the United States. In *Landscape and race in the United States*, edited by Richard Schein, 1–21. New York: Routledge.

Shabazz, Rashad. 2015. *Spatializing blackness: Architectures of confinement and black masculinity in Chicago.* Champaign: University of Illinois Press.

Spicuzza, Mary. 2019. "Milwaukee is the most racially segregated metro area in the country, Brookings report says." *Milwaukee Journal Sentinel*, January 8, 2019. https://www.jsonline.com/story/news/local/milwaukee/2019/01/08/milwaukee-most-segregated-area-country-brookings-says/2512258002/

Sweet, Joseph, Emppu Nurminen and Mirka Koro-Ljungberg. 2020. "Becoming research with shadow work: Combining artful inquiry with research-creation." *Qualitative Inquiry* 26, no. 3–4: 388–399.

Tarpley, Natasha. 1995. "Introduction: On giving testimony or the process of becoming." In *Testimony: Young African-Americans on self-discovery and black identity*, edited by Natasha Tarpley, 1–10. Boston: Beacon Press.

Thompson, Nato. 2017. *Living as form: Socially engaged art from 1991-2011.* Cambridge: MIT Press.

Trigg, Dylan. 2012. *The memory of place: A phenomenology of the uncanny.* Athens: Ohio University Press.

Tyner, James. 2006. "Urban revolutions and the spaces of black radicalism." In *Black geographies and the politics of place*, edited by Katherine McKittrick and Clyde Woods, 218–232. Toronto, Canada: Between the Lines.

Wilson, Fo. 2017. "Artist statement: Praising cabins and the black imagination." In edited by Polly Morris *Eliza's peculiar cabinet of curiosities*. Milwaukee: Bradley Family Foundation, Inc.

Woods, Clyde. 1998. *Development arrested: The blues and plantation power in the Mississippi Delta*. London: Verso.

Yaszek, Lisa. November, 2006. "Afrofuturism, science fiction, and the history of the future." *Socialism and Democracy* 20, no. 3: 41–60.

8 A poetics of opacity

Towards a new ethics of participation in gallery-based art projects with young people

Elizabeth de Freitas, Laura Trafí-Prats, David Rousell and Riikka Hohti

8.1 Introduction

This chapter elaborates on an ethics of participation in the context of creative inquiry with young people in art galleries. More specifically, the authors discuss a project called *Sensing Time*, which engaged researchers from Manchester Metropolitan University's Manifold Research Lab (the authors) and the Young Contemporaries, a youth gallery programme at the Whitworth Art Gallery in Manchester. In this project, the gallery was approached as a creative space for working collaboratively with the Young Contemporaries to cultivate new techniques for sensing the complex temporalities of contemporary art. We treated the gallery as a space for experimenting with the relational capacities of art, pedagogy and collective forms of sensing. At the start of our project, the Whitworth Art Gallery was featuring *Thick Time*, an exhibition of five complex installations by South-African contemporary artist William Kentridge.

Kentridge's exhibition had an aesthetic and political richness that was an ideal terrain for exploring the complex nature of participation. His bold architectural modification of the gallery radically transformed the museum space from an environment that traditionally conveys political neutrality and containment (Duncan 1995; Wallach 1998; Grunenberg 1999). The exhibition created a distinctive multi-sensory ecosystem of multi-channel sound, film, animation, sculptural, choreographic and architectural components with potentials to transverse and extend affectively in participant's bodies. The artworks relayed disparate sounds, chants, atonal music and polytemporal rhythms through non-linear imagery and obliquely positioned megaphones, altering the visual-sonic fabric of the space continuously. Rooms were also occupied by strangely altered antique media devices, large kinetic breathing and banging machines, which maintained a consistent pulse of rhythmic activity in direct contrast with the shifting polyrhythms of the surrounding audio and video screenings.

Our project involved a remixing of *Thick Time*, exploring the Young Contemporaries' engagement with the gallery space during six workshops and an installation. Workshops brought about 20 young people together for three hours at the gallery on each occasion. We were focused on the ways in which new media might be mobilized to alter the sense of time and participation, and how this altered form of engagement was linked to the themes and methods of Kentridge's *Thick*

DOI: 10.4324/9781003027966-11

Time. We sought to experiment with various technologies, conceptual propositions and techniques, with the aim of sensing the affective atmospheres of the
exhibition, plugging bodies into the liveliness, intensity and multi-temporality of
the artworks. We were particularly interested in the sensory dimension of gallery
experiences, noting that the senses become attuned to shifting compositions of
light, sound, space, colour, movement and form. We were also interested in how
gallery time can feel suspended, altered, slowed, quickened or transfixed as bodies become affected by the particular atmosphere that an exhibition conjures.

Our experimental techniques in the workshops included: (1) *The haptic eye*,
which involved the use of hand-held and chest-mounted video cameras as drawing tools for tracing the movement of Kentridge's art-media machines. These
videos were then cut into still images at half-second intervals, printed on paper,
then traced onto wax paper, photographed and reassembled as a form of stop-motion
video. (2) *The sensing body*, involving wearing biosensors that captured changes
in electrodermal activity, relayed via Bluetooth to a laptop. Participants moved
together through the exhibition rooms while simultaneously watching the
graphical rendering of the five biometrical responses in different colour codes
on the computer screen. Later these devices were installed as part of a concept
activation game which further disrupted the biometric signals. (3) *Sounding time*,
in which stereophonic headphones were plugged into a binaural audio-recorder
and worn as bodies circulated through the gallery space, multiplying and magni-
fying the tonalities and sonic atmospheres. (4) *Fugitive whispers*, which involved
selecting particular places in the gallery, huddling in a small group, and then
reading aloud, in a whisper, poetic texts about sensation and time. These whispers were recorded and later installed into small containers evocative of travel
and departure, themselves installed within a dark inflatable dome of silence
within the gallery.

Kentridge's *Thick time* was an interesting environment for experimenting
with these techniques of gallery sensation because the work explicitly focuses
on media-rich entanglements of collective memory, colonial history and embodiment. His installations wove together reimagined histories of colonialism and
revolution with the philosophy of art, cinema, science, and technology and identified with a practice of working with fragments. Kentridge (2014) has explained
this approach simply "as the way of going through the world; of going through
the world making sense. There is no other way. We don't have complete information, and we cannot take it in" (min. 12). Kentridge's work refuses the clarity
of a single or coherent narrative of both coloniality and revolution, and thus
produces new cultural forms through a polytemporal mix of imagery, sounds and
materials.

In this chapter, we focus on how the *Sensing Time* project entailed a distinctive
ethics of participation. We draw on the ideas of Édouard Glissant (1928–2011)
to rethink participation in terms of a poetics of opacity, showing how the project involved the collective making of a rhizomatic network of errant relays.
This approach, we argue, offers a way to conceptualize urban art projects that
might productively stray from standard gallery practices for cultivating youth
participation through a politics of identity. Our approach affirms the power of

the imagination in opening up unexpected paths for "errant" and creole forms of participation. We first discuss the context for the project, touching on the normative ethics of participation in art galleries. We then present key ideas in the Antillian philosophy of Glissant and use these concepts to analyse participatory techniques in our project. Overall, we seek to problematize the assumption that youth art projects must begin with young people revealing their positionality and knowledge, and that transparency is possible at these junctures. We advance instead a technique of participation based on a poetics of opacity, the errant and rhizomatic nature of collective forms of becoming while learning with art. In the final sections, we show how Fred Moten's (2003, 2014, 2018) work on the aesthetic excess and appositionality of black bodies supports Glissant's image of a networked urban life thriving through opaque and imperceptible relations.

8.2 Youth arts in the city

The Young Contemporaries programme in Manchester was the result of a national initiative called the Circuit project, which began in 2013 as a four-year £5 million Arts Council funded project dedicated to creating modalities of access and participation for harder-to-reach young people, via gallery-based work. Circuit involved five major galleries across the United Kingdom, including the Whitworth Art Gallery. As Sim (2019) describes, Circuit emerged as a national impetus to use local gallery spaces to improve "youth provision" at the local level, in a time when both the financial crisis and the Devolution initiative of the Coalition Government had thinned the public sector and provoked a high closure of youth centres across the country (UNISON 2018).

Circuit aimed to develop collaborations between urban young people and artists, along with projects in partnership with other local organizations in the youth sector. The gallery was perceived as a space to socialize young people and develop feelings of safety and attachment. Circuit emerged in the aftermath of the 2011 London street riots against police violence, where both government and media demanded further governance of a youth population, described by Prime Minister Cameron (2011) as lacking the proper ethics (Jensen 2018). Notably, the aim of initiatives like Circuit was to change the life-direction of harder to reach youth in a moment of social panic about the unruliness of young people. It was a move that sought to pin down youth movements, to recentralize resources, and to join galleries and youth workers, in a time of unprecedented cuts and high uncertainty. Simone (2016) claims that this sort of neo-liberal reformatting of urban space "narrows possibilities for political expression and community organizing" (198).

In the most comprehensive study of youth engagement in galleries after the UK Devolution, Sim (2019) examines differences in work methodologies between artists, gallery workers and youth workers. She writes about a possible inability of gallery workers to recognize "the uncomfortable reality that this type of work can reproduce systemic hierarchies and rituals of exclusion" (190). Sim claims that youth workers are "closer to the social and cultural background to the young people they work with" (190). These insights into the disparity between elite art cultures and diverse youth cultures, however, have triggered unfortunate forms of public engagement in many art galleries, where art has come to be that which rescues young people from their chaos, whilst affirming a narrow cultural identity.

Hickey-Moody (2013) argues that youth art programmes in cities tend to adopt categories that represent youth in pathological and deficient ways in order to justify how the arts add value and improve or engage specific demographics. These programmes do not challenge a problematic category but enact its governance. Consequently, collaborative art practice is rendered as a process of social facilitation rather than as a process of aesthetic inquiry. As Bishop (2012) notes, such distinction is important because setting a focus on the intrinsic and personal interests of the youth often involves a renunciation of the irruptive specificity of artistic practices. This renunciation further erases the dissensual and disruptive potential of the aesthetic, its capability to interrogate how the world is organized and to imagine other possible distributions of ideas, sensibilities, perspectives, spaces, embodiments (Rancière 2004).

These tensions were evident at the Whitworth in relation to our own project. The gallery leadership of the Young Contemporaries programme asked that we begin our workshops from the perspective of young people's knowledge, rather than from art and research propositions. Their assumption, echoing Sim, was that youth engagement had to begin, firstly, by rendering the young people transparent, and secondly, delivering strategies of artistic engagement that pursued clarity of identity as well as positive and non-ambiguous behaviours. We did not follow this recommendation, responding instead to what we perceived to be an aberrant *flow* of youth participation. We were working with a different approach to the ethics of participation, attempting to develop alternative participatory models. In the next section, we elaborate on this alternative, drawing on the ideas of Édouard Glissant.

8.3 Rethinking participation with Édouard Glissant

Glissant was a Francophone poet, novelist and philosopher from the Caribbean Island of Martinique who developed a distinctly archipelogic philosophy which has been described as "simultaneously Deleuzian *and* Caribbean" (Drabinski 2019, 99, emphasis in original). Glissant's poetics of opacity cultivates an onto-epistemological resistance to rendering the other transparent and legible. Opacity is the result of the accumulation of fragments from many journeys, crossings and remnants that configure colonial and postcolonial geographies. Opacity expresses a nomadic and differential view of a network of relations, as a modal process of becoming, originated through promiscuous mixings and frictions that make clarity practically impossible. Glissant describes such opaque intermixings in terms of continuous variations taking shape through "a limitless *métissage*, its elements diffracted, and its consequences unforeseeable" (1997, 34, emphasis in original).

Such imperceptibility or opacity, however, is not a deficit but an affirmation of Relation (Glissant utilizes the capital R). As Drabinski (2019) notes, this is not Relation as a correlation or a dialectic that operates to clarify or resolve, but a relational ontology that relinks fugitive memories in a rhizomatic network. Relation never pre-exists but *is* the movement of spatial, temporal and linguistic flow that brings earth, place, city and body into contact. In the special philosophical sense offered by Glissant (1997), Relation operates as an "intransitive verb" (27), in other words, as *activity without an object*. Relation cannot be correlated with the term r*elationship*, and no longer has any need to ask the normative question: "relation between what and what?" (27). A poetics of opacity attends to the lyrical remixing of the relational ontology in which life is not dependent on roots or identities (e.g. youth, gallery worker, cultural worker, artist) but lives through fragmented, unplotted, differentiating modes of being that are provisionally in contact. For Glissant, working in the Caribbean on the rich lyrical creativity of creole forms of expression, opacity was part of a political aesthetics.

But creolization is not just a way of hiding from the colonial gaze, the plantation regime and other colonial structures of control. Glissant (1997) suggests that creolization be considered an aesthetics of "baroque derangement" (91), a confluence of past classicisms, always partaking in marginality and making manifest the prior exclusions of a classical aesthetics. For Glissant, creolization is a powerful écho-monde or world-echo that performs "the integrating violence of contaminations" (91) whilst also having to "renew itself on the basis of a series of forgettings" (69). Creole-baroque techniques favour expansion over depth, open infinite dispersal over closure and diffractive lyrical relays of past "fugitive memories" (6) and "métissage" (78). Creolization is a poetic composition that emerges from the aesthetic contact, link, latency and relay of fragments, which together form a linked."totality" (17) (*totalité-monde* in the original) which is an *open processual whole*. Glissant's concept of "errantry" (18) references a kind of nomadic wandering or deviation that produces a relay, well suited to his Antillian philosophy. Without an originating explanatory story of self, we follow instead an errant diffractive movement, relaying and rerouting fragment after fragment (Drabinski 2019).

Urbanist AbouMaliq Simone (2016) affirms that in cities the right to opacity constitutes

> a potential hard to come to words … it is something that moves across territories and situations as a manoeuvre to gather, cull and distribute knowledge that cannot be pinned down. A resource to go with the 'curse' and that belongs to no one in particular.
>
> (184)

Glissant (1997) seeks this lyrical zigzag expression that distorts the rational ordering of the city with its centrifugal forces, pursuing an ethics of opaque participation that affirms both the écho-monde of contamination and the open proliferation of relays and improvisations. This achieves more than simply the refusal to comply with demands for clarity and identity: it opens onto an *ethics* of participation. Aberrant movement across the network engenders an opaque connectivity that is the condition of ethics: "widespread consent to specific opacities is the most straightforward equivalent of nonbarbarism" (194).

Glissant's Relation opens onto new forms of participation in galleries organized around an aesthetics of break, disorder, friction, fugitivity or diaspora. Rather than encapsulating young peoples' knowledge in an origin or *starting point* from which workshops might emerge, a gesture that undermines the errant entry in favour of identity, we aimed to open onto *appositionality*. Appositionality invokes a valorization of opacity, not as a rendering of things invisible or illegible, but a deviation from thinking subjectivity as origin or node. In the remaining sections of the chapter, we discuss this approach to an ethics of participation in our project at the Whitworth Gallery.

8.4 Creating new sensing bodies in the gallery

Our project took Kentridge's *Thick Time* as a space for creative experimentation with young people and the sensory conditions of the gallery. The insertion of our project within the exhibition was itself an act of creolization and non-origin, creating workshops like relays in the midst of a travelling art show before it moved to the next urban centre. We discuss here a three-channel video installation *Notes Towards a Model Opera* (2015) (Figure 8.1), collectively created by Kentridge, choreographer Dada Masilo and score writer Phillip Miller, and describe how this installation was further relayed through the participation of young people.

The moving images of *Notes Towards a Model Opera* feature a black female ballet dancer (Dada Masilo) and two black male musicians sometimes playing large brass instruments, other times carrying banners and other props. The dancing bodies of the three performers move in front of flickering images of notebooks, drawings, jottings, photographed archives, images of maps, newspaper headlines from the time of the Paris Commune, communist slogans in Chinese and English, an ink-drawn sparrow flying through the pages, and, and, and.

Figure 8.1 Stills from the video recorded in the gallery as the young man danced over Dada Masilo's dancing.

In a lecture titled *Peripheral Thinking,* Kentridge (2015) discusses how the motivation for creating *Notes Towards a Model Opera* occurred on a trip to China, contemplating the culture of the Chinese Revolution. Model operas were a theatrical genre produced between 1966–1976 that combined opera, martial arts and ballet to narrate exemplary histories of revolutionary youth. Kentridge addresses a series of what he calls *peripheral thoughts* emerging from this encounter with the model operas. He describes ballet as a bourgeoise genre that emerged in European urban centres at the end of the nineteenth century and inexplicably moved to the Union of Soviet Socialist Republics (USSR) and later to China to tell revolutionary stories. Kentridge speaks of the absurdity and beauty enacted by ballerinas dressed in revolutionary gear, holding rifles and shooting at Japanese soldiers, while performing technically audacious ballet movements. He wonders about the epic, exaggerated style—in Glissant's terms, the operas combine fragments of European Bourgeois classicism with revolutionary futurism.

In collaboration with Dada Masilo, a contemporary South African dancer and choreographer internationally known for decolonizing and queering classic ballet pieces, Kentridge began a further exploration of the model operas. They developed dance improvisations incorporating the sounds of 1950s colonial dance bands, which evolved into the current artwork. Considering all its fragments and nuances, *Notes Towards a Model Opera* delivers an aesthetically, geographically and historically thick landscape of multiple scales and lifeways set in variation and free association that carries and cultivates opaqueness. Its fragmentary creolization of classical aesthetic forms became the conditions for an Antillian relay, a true métissage in the sense proposed by Glissant.

Now consider this vignette from our *Sensing Time* workshop: A young black man dances into the exhibition room equipped with a pair of stereophonic headphones connected to a binaural sound recorder. His movements coordinate with a sonic scape, enabled by the technology that he is wearing and that others in the room cannot hear. His dance has been captured by chance on a GoPro camera mounted on a chest harness of another young person, also passing through this room. His individuated dance joins the movement of the ensemble, in which disparate moments/bodies/languages (dance and image) stick together and co-compose. The choreographer Dada Masilo finds an unexpected dance partner, a relay linkage between her performance and the errant nomadism of this young black participant. This rootless production of relays begins to form an open rhizomatic network, the expression of an opaque ethics of participation. As Glissant (1997) suggests, "it is the network that expresses the ethics" (193).

The room is filled with the soundtrack of *Notes Towards a Model Opera*, a version of the Socialist anthem *The International*, transformed by the sounds of a brass marching band, replete with offbeat and African polymetric sounds. The mixing of different tonalities and rhythms evokes a métissage of cultural formations that are mutually deterritorializing (Glissant 1997); *The International* becomes more like a dance tune than a communist party anthem. As an anticolonial smuggler who enters a space where almost everyone is watching and talking, the young man dances and shuffles sideways towards a group of friends, smiling, shaking hips and opening arms. His dance affirms the inseparability of African music, dance and the sociality of black life (Moten 2018) by provisionally disrupting the museum's civilizatory rituals (Duncan 1995) which typically try to maintain modal separations between visuality, hapticity and orality (Campt 2017). The young man's dance is an enfleshed crossing of such borders, an interplay of sensual, social and aesthetic ensembles. His dance is conditioned by the environment, where the body folds towards the environment, and recomposes in fusion with it (Manning 2012).

The joy and easiness of this improvised dance contrast with Dada Masilo's choreographic convolutedness. Dressed in military apparel and ballet shoes Masilo's dance is a more explicit écho-monde of colonial violence. She moves with an uncomfortable material and cultural hybridity; on the screen, her poised movements are folded into an exhaustion of fragmentary sounds, words, drawings, journal headlines, notebook pages, archival documents, languages, geographies, temporal scales, saturated tonalities. *En pointe* she holds two red flags and two automated old-fashioned pistols. *En pointe* she carries a rifle hanging across her torso. *En pointe* she moves both hands in syncopated gestures, and shoots, shoots, shoots, as *The International* continues playing with the loud beats of a cowbell and the heterogeneity of images rapidly flickering in the background.

We do not know the origin or motivation of the young man's dancing in composition with these moving images. We use this vignette to show how participation involved another relay in the dense composition of the artwork, where different chords and rhythms are part of an ongoing, evolving melody (Simone 2019) and different dancing modes (Masilo's and the young man's) are part of one choreography of Relation. The young man's participation remains opaque in the sense that it neither coheres with the concerns of the artwork, nor with the behaviour of his friends, nor with the intentions of our research, but inserts a fugitive métissage, an écho-monde, a creole-baroque derangement, and ultimately a poetic improvisation that opens up the network. We note how the use of binaural sensor technologies altered the atmospheric texture of his participation, allowing for a kind of technogenesis through the use of the new media and the matrix of the sensing body in movement (Manning 2012). For Manning, technogenesis constitutes a dynamic becoming or transduction of the sensing body in movement through technicity, a process that expands and opens the body-as-event into alternative networks of relation within an emergent environment. But this vignette was just one minor relay in the unfolding of *Sensing Time*. In the next section, we discuss the project in more detail.

8.5 Cultivating the logic of the relay

The *Sensing Time* project moved through a series of participative relays in the Whitworth Gallery, firstly through workshops with the Young Contemporaries, and secondly through an interactive exhibition several months later. Our aim during the exhibition was to continue to widen the networks of participation through a series of immersive artworks. Each of these artworks involved remixing, sampling, reanimating, projecting and drawing on the techniques of experimentation used during the workshops. The exhibition opened at the Whitworth

Art Gallery in May 2019 with the title *Re-mixing Thick Time* (See Figure 8.2). Participants from the Young Contemporaries attended, along with numerous colleagues and contacts, and members of the general public.

Onto one wall of the Whitworth Art Gallery's Grand Hall, we projected a film titled *The Missing Half Second* and arranged a series of beach chairs for viewing. The film was created using short videos made at the Kentridge show that were later split into 120 half-second still images. These images were printed and brought back to a subsequent workshop, where they were redrawn by the research team and the members of the young contemporaries, as black and white drawings on semi-transparent paper. These 120 redrawn images were then recomposed in layers and reanimated as a moving image, smudging the temporality of the event, to foreground the rich multiplicity and ghosting of the present moment. The film explores the idea of a missing temporal interval between sensation and conscious perception (Massumi 2002), by continuously interleaving and superimposing the disparate images.

The film is approximately 9 minutes in duration and can be viewed online at https://vimeo.com/326813533. The animation documents a participant as she herself videotaped a strange archaic set of gears placed on top of a large wooden tripod mounted by Kentridge in the art gallery, moving her own camera in a circular fashion to animate the potential movements of this antique machine. In many frames of the film, we find that the gears of Kentridge's machine, the lens of the camera, and the face, arms, and hands of the young woman become indistinct (see Figure 8.3, above). They begin to circulate as an atmospheric body—the

Figure 8.2 General view of the *Remixing Thick Time* Exhibition. From foreground to background, we can see the installations: *A game of Conceptual Activations, Fugitive Whispers* and *The Missing Half Second.*

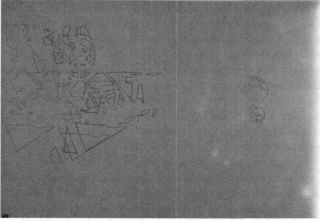

Figures 8.3 The Missing Half Second. Animated video. Still images #40–45 layered (above) and still images # 66–71 layered (below).

young woman's body disappears and reappears, slowly transformed into fragments of gears, fingers, wheels, etc. (Figure 8.3, below).

Each of the original 120 drawn stills also lay along the gallery wall, on benches. These were another shadow and echo of the relay. The process of sampling and remixing media rendered a sense of being caught up in a maker's loop, which kept us switching back and forth between the still image and the moving image. The moving image was relayed and proliferated to such an extent, folded into itself, that it seemed to become the substance of an entirely different media, something more atmospheric, as though we had contaminated the gallery with the layerings of these dynamic flowing images. The *Missing Half-Second* was further relayed through a reprogrammable sound system, which invited audiences to remix samples of sounds recorded by young people in the gallery, including the slamming sound of old machinery and the singing sounds of South African colonial music.

The Missing Half Second exemplifies Glissant's notion of *relay* as rhizomatic spread. The relay avoids the literal replication of any fragment because it renounces the assumption that there is one beginning from which the fragment stems. The relay is a constant detour, mix, and complication of elements through the movement-force of Relation (Davies 2019; Drabinski 2019). This is expressed through Glissant's figuration of the *relay agent* as that which moves with Relation as an *open* totality (tout-monde), maintaining a durational expanse of iterative variation. In this respect, the relay both emerges from and disappears into the opacity of Relation. Glissant (1997) contrasts relay agents with what he terms *flash agents*, which attempt to illuminate and fix identity in a flash of transparency akin to the flashbulb of a journalist's camera. This flash arrests the event in the process of its unfolding, in an attempt to capture, interpret and represent the world in a single, universalizing gesture.

In another installation called *Fugitive whispers*, the recorded whispers from the Young Contemporaries workshops quietly emanated from time-worn travel containers inside a pitch-black dome occupying the middle of the gallery Grand Hall (See Figure 8.4). Visitors crawled into the dome with flashlights, straining to hear these whispers. We wanted to emphasize how whispers resist the urge to broadcast and refuse to fill the space, enticing bodies to come closer and stretch their senses beyond what might feel comfortable. Whispering in the enclosed dome was a politics of counter-speech—quiet, intimate, proximal, close, secretive, opaque, not to be heard by all. The whispered poetic texts spoke of the precarity of sensation, reminding listeners of the secretive lyrical voice.

As an event, the *Re-mixing Thick Time* exhibition and its series of installations sought to circulate and exchange a repertoire of skills, senses and sensibilities forged through the empirical, aesthetic and situated theoretical possibilities generated in propositional works of art (Corsín Jiménez 2014). Perhaps, the participants carried these modes of attention further afield, transiting across the city (Simone 2016). Perhaps, the social, aesthetic and technologic gallery conditions catalyzed more errant relays, suspending the neo-liberal reformatting of urban space as an orderly gridlock (Thrift 2014; Simone 2019). We don't know. There was—by design—no proof of this: "What best emerges from Relation is what one senses... Relation cannot be 'proved' because its totality is not approachable" (Glissant 1997, 174).

8.6 A durational ethics

As the project evolved, we became increasingly aware of our discontinuous and fragile relations with the Young Contemporaries. As an open-door programme, young people who came to the first workshop did not necessarily return to later workshops. In many cases, young people came to our workshops with no prior experience of the Young Contemporaries and were invited to join in with complex theoretical and technical propositions about art and sensation. This rendered any normative sense of continuity or sustained participation through attendance questionable. For example, the young man dancing into the gallery did not come back to further sessions. Later, we learned that he was one of the

Figure 8.4 A young person listening for his voice in the *Fugitive Whispers* installation.

original participants in the Circuit project back in 2012 and had since grown out of the Young Contemporaries age range of 16–25 but still came back to hang out when he felt like it. He danced into the gallery and didn't come back, performing a relay, a rerouting, affirming the opacity of the event, a "gesture of giving-on-and-with" (Glissant 1997, 192).

This and other anecdotes made us further examine the ethics of the project. While the disjunctive and errant flow of young people's participation made our project experimental, it also raised issues about equity and value. As researchers we walked away from each workshop with fascinating insights, artefacts and data rendered through youth participation, but what were the young people taking away? Were we sampling and appropriating young people's experience and utilising it to advance our own academic and artistic ambitions? One of us wrote this reflection on the experience of talking with a young person after he listened to his whispered texts in the *Remixing Thick Time* exhibition:

> "Did you hear yourself?" I asked the young man who had participated in the workshop where we recorded the whispered texts. "Yes", he replies, smiling as he scrambles out from the dome, "but I didn't imagine it this way". What

did he imagine? He who trusted us and gave his consent to become a participant. Did we steal his whispers from him during that workshop?

The anecdote reminds us of the powerful role that surprise and imagination play in generative art. An active relay links the actual localized sensation in the dome to the errant imagination, to the infinite speculative reach of the imaginary and the poetic. Perhaps an ethics of participation that is committed to opacity must move with a certain aesthetic, a certain way of sharing and sensibility that permit many relays to occur, growing multiplicity and pluralism (Davies 2019). The whispering workshop involved an intensely alienating series of intimate huddles where strangers came together in a singular moment of poetic language, within the cold confines of the gallery. Participation through whispering involved a productive amount of discomfort and the generative weirdness of the "strange stranger" (Morton 2016). There was no transparency; whispering made that fact clear. As Glissant suggests, in Relation there is no depth of essence to be illuminated and grasped in the transparency of the other; self and other are rendered opaque to one another, without essential, originary or universal filiation.

> I thus am able to conceive of the opacity of the other for me, without reproach for my opacity for him. To feel in solidarity with him or to build with him or like what he does, it is not necessary for me to grasp him.
>
> (Glissant 1997, 193)

Yet it is precisely in this opacity of Relation that alternative forms of togetherness might be forged, if not grasped. And this cannot be forged if stolen whispers and fugitive memories are only meant to reveal an originating truth about self or identity.

The *Sensing Time* project operated with temporal, spatial, material and pedagogic limitations due to the reality of being visiting artists-researchers in a gallery youth programme for only six three-hour sessions with different groups of young people every session. Our regular conversations as researchers reflected on a certain gap between the potential for relation that *Sensing Time* proposed (as a series of concepts, techniques and technologies) and a concern that our workshops did not allow for the slow modes of mutual exchange required for Glissant's active relay. This is a brief fragment from our April 2019 research notes, when we discussed the ethical complications of participation as the project was unfolding:

- Can we call it collaboration? Or is it a different relation?
- We can't pretend that there is an equal investment.
- Are the young people our advisors? We go there with our ideas—forcing them to witness something…
- We have to resist romantic language.
- Invitation, proposition, power relation, who profits? Who puts the names in publications?
- It's a production. We produce a work of art together.
- Who is "we"?

- Advisors, counsellors: We have gone to them.
- We catalyze something, a flow of energy, affect. Vulnerability, the charge that they bring is a certain vulnerability.
- We go there to learn, we learn from the relation, how ideas move.
- Sometimes you feel selfish…
- We are dealing with complex things that need more time.
- What is "healthy participation"? Complexifying in participation, empty futures.
- It's more about co-production of an event, not individual relations.
- Exchanges of epigenetic particle-signs, event-particles.
- Infra-individual—there is a kind of molecular poetics going on.
- An ecology of dependency relations and contamination in dependencies.
- Life has stepped in, interfered with our beautiful plans.
- Something was there before we came.
- Not ethnography but syncretism, you are participating in each other's worlds…
- If it's an ecology then we are the monsters, the big and powerful ones in the network. They are another kind of being in this ecology.
- Symbiotic relationships can be mutually beneficial, parasitic or predatory. The future of the relationship is uncertain.
- The young people carry on to other ecologies, they have given something, they go somewhere potentially to be monsters themselves.
- An ecology of participation.

We emphasize here the stated concern for how workshop participation might entail an empty or uncertain future, and how a focus on new media and the senses might be considered a molecular poetics in which event-particles contaminate and transform the atmosphere. We were painfully aware that an ecological attention to symbiotic relationships and syncretism must reckon with power differentials and the potential of monsters—and monstrous imaginations—dominating the dependency relations. We were the researchers, structuring the workshops. We directed the activity by introducing new sensor technologies and new digital media, deliberately linking up with Kentridge's monstrous media machines as they relayed images and sounds of imperial invasion and violence. Media operate as flash agents when they freeze the relation in coercive subjection, and broadcast nationalistic anthems, or amplify the speed of connectivity for the purposes of control (Curto 2016). For this reason if no other, it seemed urgent that we consider this project in Antillian terms of creolization, relay and écho-monde. An active relay further difference/tiates the network into the space of poetics and creative creolization. The concept of *rerouting* in Glissant (1997) (translated from French *détournement*) emphasizes the continuous movement of knowledge within relationality, no matter how arrow-like the relation may appear. For Glissant, relay is not a literal repetition but performs a certain detour, a baroque derangement. The ecology of participation moves as an open whole: The world as totality (tout-monde). Harney and Moten (2013) also think with the concept of rerouting, affirming the work of art is where Blackness "discovers the re-routing" (50). They help us make sense of the specificity of the Kentridge show and its appearance in Manchester, elaborating how the work of art "animates a range of social chromaticisms", and enables mutation through "imposed", "self-generated" and "stolen" relays.

Moten (2003) argues that works of art can lead to performances-in-objection, but there is also a speculative excess that escapes. Likewise, acts of whispering and recording can lead to different imaginaries about how those whispered texts might surreptitiously tiptoe into an unknown future. "I did not imagine it this way" invokes imaginaries that are not necessarily revealed in the work, but simply *are* its poetic opacity, its unscripted futurity, its divergent expression. Harney and Moten (2013, 51) write about artistic techniques as carrying a performative aspect that permit an escape or a flight that may feel like "a constant economy of misrecognition"; misrecognition because the escape can be neither articulated in words nor rationalized. No doubt we participated in an economy of misrecognition, where we were trying to *grasp* and convey something. But an ethics and poetics of opacity help us to see how our grasping gave way to a different kind of gesture altogether, "the gesture of giving-on-and-with that opens finally on totality" (Glissant 1997 192).

8.7 Conclusion: Relation in the city

A poetics of opacity demands a commitment to the fragmentary form, which transforms the conditions of representation in two ways (Drabinski 2019). Firstly, an aesthetics of the fragmentary eschews metanarrative and filiation, making it impossible to tie subjectivity to a single origin or root, and secondly, it undermines "a regulative idea of transparency" (139). Rather than seeing fragmentation as a source of isolation, Glissant sees it as a vital force of nomadic life, a force that circumvents demands for an atavistic, authentic and original self or positionality. For Glissant (1997), an ethics of opacity involves a commitment to the intransitive totality of Relation and its expression through imagination. The imagination relays a thousand potential futures, of which only one will be actualized. There is no unitary direction to be taken from Relation because it is an "open totality evolving upon itself" (192). The force of poesis or artfulness has its own singularity of movement in this respect, to the extent that the work of art reroutes an ecology of participation in the deviant force of its own becoming, forming a collective but dispersed "sensing body" (Manning 2012).

Considering the artwork as an appositional rerouting has implications in relation to Sim's (2019) argument about the symbolic violence involved in art gallery programmes for youth engagement. It suggests that alongside such symbolic violence there may be an aesthetic excess that opens up alternative modes of sociality and fugitive forms of participation and improvisation. This is a poetics of opacity that engages in processes of re-materialization outside the homogeneity and exclusions of architectural design and city planning. For Moten (2014), the aesthetics of appositionality produces a thinking at the edge of the city as built space, in which it is possible "to think and inhabit an architecture whose rematerialization makes it an architecture outside architecture" (167). This work opens onto the inchoate complexity of civic participation, sensing the fugitive, illegible, inconsistent, hard-to-notice performances that escape from neoliberal governance and planning. Moten helps bring Glissant's aesthetics of appositional relation to a reading of the city and its aesthetic dimension, composed by errant movements and transitory linkages. Simone (2019) also writes of an economy of relation in the city where people are not identified and incorporated but instead indulge in "tasks with their own time … detached from ins and outs … with an openness of what the exchange could entail" (44). Refusals and resistances move across the city, wherever youth are called to spend time whilst they imagine errant paths and other futures. The *Sensing time* project used diverse techniques of participation so as to cultivate these kinds of refusals as creative manoeuvres, responsive to the aesthetic and atmospheric conditions of the gallery.

References

Bishop, Claire. 2012. *Artificial hells: Participatory art and the politics of spectatorship*. London: Verso.

Cameron, David. 2011, 11 August. "*PM statement on violence in England*". [Speech] London: House of Commons. https://www.gov.uk/government/speeches/pm-statement-on-violence-in-england.

Campt, Tina. 2017. *Listening to images*. Durham, NC: Duke University Press.

Corsín Jiménez, Alberto. 2014. "Auto-construction Redux: The city as method". *Cultural Anthropology* 32, no. 3: 450–478. https://doi.org/10.14506/ca32.3.09

Curto, Roxanna Nydia. 2016. *Inter-Tech(s): Colonialism and the question of technology in francophone literature*. Charlottesville, VI: University of Virginia Press.

Davies, Benjamin. 2019. "The politics of Édouard Glissant's right to opacity". *The CLR James Journal*, 25, no. 1/2: 59–70. (first published online) https://doi.org/10.5840/clrjames2019121763.

Drabinski, John. 2019. *E. Glissant and the middle passage: Philosophy, beginning, abyss*. Minneapolis: University of Minnesota Press.

Duncan, Carol. 1995. *Civilizing rituals: Inside public art museums*. New York. Routledge.

Glissant, Édouard. 1997. *Poetics of relation*. Translated by Betsy Wing. Ann Arbour, MI: University of Michigan Press.

Grunenberg, Christoph. 1999. "The modern art museum". In *Contemporary cultures of display*, edited by Emma Baker, 26–49. New Haven, CN: Yale University Press and Open University.

Harney, Stefano and Moten, Fred. 2013. *The undercommons: Fugitive planning and black study*. New York: Minor Compositions.

Hickey-Moody, Anna. 2013. *Youth, arts and education: Reassembling subjectivity through affect*. Abingdon, UK and New York: Routledge.

Jensen, Tracey. 2018. *Parenting the crisis: The cultural politics of parent-blame*. Bristol, UK: Policy Press.

Kentridge, William. 2014 "How to make sense of the word". Interview. Filmed [2014]. YouTube video, 30:24:07. https://www.youtube.com/watch?v=G11wOmxoJ6U

Kentridge, William. 2015 "*Peripheral thinking*". [Lecture] New Haven, CN: Yale University Art Gallery. YouTube video, 1:00:23. https://www.youtube.com/watch?v=79FuROwzRvs

Manning, Erin. 2012. *Relationscapes. Movement, art, philosophy*. Cambridge, MA: MIT Press.

Massumi, Brian. 2002. *Parables for the virtual: Movement, affect, sensation*. Durham: Duke University Press.

Morton, Timothy. 2016. *Dark ecology: For a logic of future coexistence.* New York: Columbia University Press.

Moten, Fred. 2003. *In the break: The aesthetics of the black radical tradition.* Minneapolis: University of Minnesota Press.

Moten, Fred. 2014 "Collective head". *Women & Performance: A Journal of Feminist Theory* 26, no. 2–3: 162–171.

Moten, Fred. 2018. *The universal machine.* Durham and London: Duke University Press,.

Rancière, Jacques. 2004. *The politics of aesthetics.* London: Bloomsbury.

Sim, Nicola. 2019. *Youth work, galleries and the politics of partnership.* Cham, Switzerland: Palgrave.

Simone, AbdouMaliq. 2016 "Urbanity and generic blackness". *Theory, Culture and Society* 33, no. 7–8: 183–203. https://doi.org/10.1177/0263276416636203.

Simone, AbdouMaliq. 2019. *Improvised lives: Rhythms of endurance on an urban south.* Cambridge, UK: Polity Press.

Thrift, Nigel. 2014 "The 'sentient' city and what it may portend". *Big Data and Society* April–June 1, no. 1: 1–21. https://doi.org/10.1177/2053951714532241.

UNISON. 2018. "Axing millions from youth work puts futures at risk, says UNISON" (December). https://www.unison.org.uk/news/press-release/2018/12/axing-millions-youth-work-puts-futures-risk-says-unison/ Axing millions from youth work puts futures at risk, says UNISON.

Wallach, Alan. 1998. *Exhibiting contradiction: Essays on the art museum in the United States.* Armhest, MA: University of Massachusetts Press.

Epilogue

The remaking of collective life in (post)pandemic times

Laura Trafí-Prats and Aurelio Castro-Varela

We began the edition of this book in 2019 with no glimpse of a global pandemic in our then conceivable horizons. We end it with the event of Covid-19 still unfolding and with no certainty of what will be its indelible matterings. Then, it appeared that giving some space to thinking with the present conditions made sense for a book whose philosophy embraces a speculative approach to arts-based research and that also aims to generate collective imaginaries beyond the current inhabitability of neoliberal cities. As we wrote, edited and rewrote the book chapters, we witnessed how the emergence, management and effects of Covid-19 intensified such inhabitability. Despite initial claims describing the pandemic as a *great leveller* (Kapadia and Sirsikar 2020), we soon learned that Covid-19 compounded with ongoing forms of violence connected to capitalism (Butler 2020), austerity (Lancione and Simone 2020a), abstraction (Simone 2020), racialization (Ferreira da Silva 2020) and the destruction of habitats in the Anthropocene (Haraway 2020). As such, Covid-19 has undoubtedly contributed to the intensification of structural differences.

While we do not seek to generate any definitive examination of the present moment, we try to engage with some thoughts and gestures that have been articulated through the first year of the pandemic in the social sciences, the arts and philosophy. We set them in tentative conversation with key arguments that we have put forward in the book. In this respect, the impossibility of separating where Covid-19 starts and where Western liberal capitalism ends resonates with one of the central motifs orienting our book. This is the interest in approaching all bio-social phenomena in less anthropocentric ways and with the recognition that all phenomena interact in complex affective and agential ecologies (Bignall and Braidotti 2019). We think that participative arts-based research in the city is well equipped to lead this affective and ecological turn in social research. This is so because we see in the experimentation with techniques drawn from architecture, digital media and the occurrent arts, ways of opening subjectivity to the movement, sensation and becoming of a world outside (Massumi 2002; Ellsworth 2005). In pandemic times, participatory arts-based research could function as a space from where to devise practices that break down with the concept of human exceptionalism, which has been so present in the science of the pandemic, by recognizing and speculatively engaging with Covid-19's situated environmental entanglements.

DOI: 10.4324/9781003027966-12

Through the pandemic, we have witnessed how human lives have been featured as an abstraction made of numbers, lines, curves, tables, predictive mathematical models and lists of positive cases and deaths internationally, per country, and per postcode. As Ferreira da Silva (2020) notes, such processes of abstraction have worked to think the pandemic through generalizations which have occluded singularity, the singularity of every life. Such quantitative abstractions have operated compounded with other existing abstractions derived from capitalist colonial systems. Terms like *underlying conditions* have been used to model and describe the likelihood of being contaminated and at higher risk of dying, without revealing that organismic, economic and spatial conditions of disadvantage are connected to historically situated processes of racialization (Ferreira da Silva 2020) and austerity (Lancione and Simone 2020a). In such context, alternative and more accountable knowledge practices of measurement, representation and mapping connections between life and death seem an important space in which participatory arts-based research can engage in creating more singular experiences.

We have lived daily with a convincing performance of science as an objective, nonsituated, disembodied, male knowledge aiming to control and combat the virus (Haraway 1988). The opportunity that Covid-19 has offered to rethink ourselves as multispecies bodies, made mostly of bacteria, fungi and the like (Haraway 2008) has been undervalued in favour of mighty humans fighting a war against the virus and flattening the curve. The virus per se, along with animals and environments, have been constantly featured as secondary forms of life, while the focus has been set on humanization through numbers (Ferreira da Silva 2020).

In contrast with this, the World Health Organization (2021, 83) has described Covid-19 as a virus with "a zoonotic origin", although "the specific route of transmission from natural reservoirs to humans remains unclear". It is possible that the tracing of such route may never be fully established. However, one thing that is well known is that the destruction of animal habitats marking the Anthropocene puts animals and humans more in contact and increases the risk of a zoonotic disease being transferred to humans. As Price (2020, 776) has remarked, "SARS, bird flu, MERS, Ebola, Zika, and Nipah have all emerged from animals".

The dual organization of Covid-19's public discourse around humanization on one side, and naturalization through the public prominence of immunological research on the other, has constituted an exemplary enactment of Whiteheads' (1964) bifurcation of nature. Giving human bodies primary qualities or underlying conditions that do not belong to them is a way of separating humans from their entanglements with the world. In turn, this has contributed to rendering nature as a dull and objectifiable place. Nature is spoken as a place where scientists gather evidence from natural reservoirs, sequence the virus, study its genome, establish clusters of transmission, etc. But as Latour (2014) remarks, "what is important to remember is that the bifurcation is unfair (...) to the human and social side as well as to the nonhuman or 'natural' side" (98). On the nonhuman side, cattle have been intensively tested, and some have been sacrificed; other animals have been manipulated in labs as testing bodies for the development of vaccines; environments have been vacated and sanitized, and circulation remapped and micro-managed, and air systems have been tuned up for air renewal and flow. Many of these interventions that in principle are to save lives create drastic separations between humans, animals and the environment. However, their relation to long-term environmental goals remains unclear if not inexistent (OECD 2021).

Working with the assumption that we feel before we identify (Massumi 2002; Ellsworth 2005), our book has foregrounded sensation over cognition, proposing an affect-based approach to urban life. By focusing on affect, we have argued that when we feel, we do not feel alone but with other entities that we enter in relation with (Shaviro 2012). Resonating with this argument, Haraway (2020) has characterized life in lockdown in terms of a sensational ecology that materializes differential modes of inhabitancy, sensitivity and connections with other multispecies that matter in how we live and die in pandemic. She reflects through several stories percolating in her household as she sheltered in place. One concerns her husband's students serving sentences in a state prison and the written correspondence that they exchange. For these students, the lockdown has involved further isolation in individual cells and less than proper access to light and environmental sounds like birds, which with other animals, have been purposely deterred to approach the prison during the pandemic. Without the sounds evoking life outside the prison, the students describe an overwhelmingly quiet atmosphere only occupied by the cries and screams of other inmates feeling their lives decomposing. Haraway speaks also of one of her graduate students' children, who have been deprived of outdoor play, face-to-face socialization with friends and access to a classroom. Their life has been sensorially reduced to spending hours with the sounds and images projected by Zoom and TikTok through a phone screen, while left home alone during the hours that their mother works in the service industry. Haraway mentions too the 21 times that George Floyd uttered "I can't breathe" before he died asphyxiated under the knee of police Derek Chauvin. She contrasts the despicable and ominous killing (and thus silencing) of Floyd to her daily activity, where being in silence carries the privilege of listening to the sounds in her garden inhabited by luscious plants, trees, hens, birds and bees. The (human) stillness of the lockdown brings to the senses both the thriving and declining forms of multispecies sociability that emerged in lockdown. On the declining side, Haraway mentions the sounds of birds falling from the sky and hitting the ground exhausted and food-deprived. Their deaths appear caused by the rerouting of their migratory journeys generated by California's wildfires and the loss of food resources. Haraway describes these sensuous and sonic landscapes as the *Phonocene*, which possibly is an extension of the concept of *Chthulucene*, Haraway's (2016) more muddled, sensuous, tentacular alternative to the Anthropocene. With *Phonocene*, Haraway highlights sound, silence and acts of listening as ecological, aesthetic and political practices, noting that a planetary crisis like Covid-19 constitutes a deep shock in the sensorium. By deliberating on how relations of silence and sound matter in pandemic times, she highlights the alliances, interdependences and processes of becoming with others that got suddenly lost, interrupted and reconfigured in the lockdown.

Echoing Haraway's work of thinking pandemic times through the threading of complex stories that matter new ways of making connections (Haraway 2016), in our book we have argued for a model of inquiry that is *constructive* (Stengers 2002), where knowledge is not out there waiting to be discovered but relative to attuning to the qualities of experience (Shaviro 2012). This resonates with Price's (2020) affirmation that Covid-19 is a phenomenon that is not there just to be fixed and then move on into a *semblance of normal* in the words of PM Boris Johnson (Behr 2021). Covid-19 is to be *lived with* since it is now central to the becoming of life on earth. Working with the understanding that the knowledge of Covid-19 is relative to experience requires a recognition that agency, rather than depending on a knowing subject, resides in a complex world of material entanglements that are agentic without necessarily involving humans (Barad 2003; Price 2020). Bignall and Braidotti (2019) note that recognizing other forms of influence in activity beyond human action is central in creating an ethical space that opens to posthuman frameworks made of heterogeneous compositions of materials, affects, desires (see also Deleuze and Guattari 1987). In relation to this idea, it is important to consider how the safety protocols of Covid-19 have eroded a feel for publicness connected to collective life in the street. Publicness carries elements of sensuousness, proximity, desire and unexpected encounters that are essential to open up to the heterogeneous posthuman frameworks of relation noted above.

Not surprisingly, notions like self-isolation, sheltering-in-place or social distance central to Covid-19 provisions have ignited discussions around publicness and collective life. On the one hand, there has been enthusiasm about modalities of public life that have become prominent during the pandemic, including the growth of community organized networks of mutual aid (Solnit 2020). On the other hand, Covid-19 has rekindled the debate about what type of public life we want (Satvrides 2020), which includes reflections around whether solidarity is sufficient, or there is a need for spaces of collective discussion, experimentation and re-materialization (Lancione and Simone 2020b). This reconnects with pre-Covid-19 times linked to austerity policies and the movement of auto-construction and infrastructuring (Corsín Jiménez 2014). It resonates too with questions posed in our introduction around the right to the city and forms of collective life which are not based on the need to secure rights from the state (Larkin 2013) but seek to generate languages, aesthetics and ways of being together alternative to Western liberal ideas of subjectivity (Berlant 2016).

With a focus on cities of the Global South, Simone (2020) discusses how Covid-19 provisions were applied expecting totalizing effects but with indifference to the urban lifeworlds of the Global South. Measures such as social distance, shelter-in-place, washing hands and working from home have been rendered as dependent on individuals' duty and responsibility, ignoring that these measures are completely relative to infrastructural conditions. Simone (2020) writes about how in South African townships, individuals from more than one family or group share the same roof and sleep together in small rooms. Water and electricity are not always ensured, in a way that washing hands and keeping them clean constitutes a real challenge. Many depend on daily food intake given by organizations or members of the community who play the role of feeding and caring for those who do not have perhaps the time and resources to feed themselves. The mere initiative of going to buy some food at the nearest market may involve several intermediary steps, contacts and movement that trump any well-intentioned safety measures.

However, as Simone (2020) argues, rather than seeing these as deficit environments that indisputably will be ravaged by the effects of the virus, it is important to learn from the practices of friendship, mutuality, exchange and invention developed in Southern communities, and how these create infrastructures that protect and care. For example, thanks to different practices of counter-information, skills exchange, door-to-door help and activism, some favelas in Rio and Sao Paulo kept their infection rates lower than their cities average (Béhague et al. 2021; see also Satvrides 2020). Such daily, flexible practices are radical examples of collective forms of responding and attuning to macro and micro transformations in the environment. Before they got hit hard with Covid-19, Southern urbanisms were already full of examples of exchange and interdependence formed and cultivated as ways to cope with compounded structural crises (Simone 2020). This is relevant when examined from the perspective of our book's interest in participatory arts-based research practices in the city because the forms of assembling collective life infrastructures in Southern urbanisms offer important cues for how to reorganize publicness in the cities of the West. We may need to stay inside to save lives. However, for our lives to go on, we need ways of attending to our surroundings so as to invent modes to space out and reinhabit the world as we criss-cross paths with others (Lancione and Simone 2020b).

Drawing on the work of artists Tomás Saraceno and Rafael Lozano-Hemmer (also, McCormack 2018; de Freitas, Rousell and Jäger 2020), we have discussed how participatory art-based practices can invent technological-spatial-bodied infrastructures that extend capacities for attuning to atmospheric conditions. Covid-19 is an airborne virus that propagates through air and breath. As such, it is a phenomenon that has clearly highlighted the importance of how atmospheres shape social life and that sensing environmental conditions with other bodies becomes a central and critical activity in the times we face. Arts-based research projects can envision infrastructures that invite and support communal experiences of active sensing and attuning to atmospheric and territorial conditions. This can contribute to ongoing projects and conversations with existing networks of solidarity and auto-construction that can enhance and protect life in situated communities, resisting neoliberal policies that propagate risk and inhabitability.

This idea of building new forms of undisciplined collective life resonates with how Roy (2020) in the early months of the pandemic noted that Covid-19, like other historic pandemics, could become a portal from where to rethink the world beyond capitalist forms of oppression, ideology and exploitation. Few months later, Butler (2020) extended this argument by affirming that rethinking the world outside capitalism's vectors of power and oppression would involve a recognition of the interdependency of all forms of life and the establishment of conditions to operate with radical forms of equality needed for a more liveable life. We suggest that Roy's and Butler's ideas, as well as any other argument about re-articulating and recomposing collective life, should be pursued with social research, philosophy and the arts at the centre for thinking and enacting more than just ways of living and dying in pandemic times, and also for recovering our bodies' intimacy usurped by the pandemic's positivist discourse.

In the book, we have argued that exposure to and participation in art-infused environments nurtures a rethinking and redoing of collective life. As Ellsworth (2005) has noted, such subjective remaking is continuous with a remaking of the world. The categorization of art organizations as unessential services and their consequent closure during the pandemic has constituted a massive loss in potential for rethinking collective life. This is so because it is through radical forms of interdisciplinarity between the arts, sciences and humanities that other forms of *making kin* (Haraway 2016) and radical interdependency (Butler 2020) can be imagined and experimented.

Although projects and activities had to be postponed, there have been efforts in the arts to generate spaces of publicness. Such publicness has not necessarily involved the presence of bodies in the streets (although this also happened in remarkable ways in the aftermath of George Floyd's killing), but it has reaffirmed the notion of *urban agency* (Amin and Thrift 2017) that we discussed in the Introduction. There, we noted that urban agency underpins the assembled character of urban life, where bodies function in arrangements of things, technologies, spaces that allow certain forms of social experience to take place. Thus, in experimenting with compositions of space, infrastructure and aesthetic interventions, some art projects have sensuously re-territorialized space and re-singularized the experience amid the ongoing abstractions on Covid-19. One early example of this was delivered by digital artist Robin Bell, who projected images extracted from the Covid Memorial, a digital archive that collects and honours the memory of those deceased by Covid-19, on buildings in a residential area of Washington DC. People sheltering in their apartments could see and read the projections from the windows of their homes, besides following the event on Twitter. The projections occurred in April 2020, when President Trump was actively belittling the pandemic, not assuming responsibility for its mismanagement and avoiding any reference to deaths.

Months later in November 2020, artist Nick Cave created a mural with the message "Truth to be told" made of black vinyl letters measuring 21-feet high and warped across the front façade of the Jack Shainman Gallery's outpost in the Hudson Valley village of Kinderhook. Cave's artwork followed the police killing of George Floyd. It has been described as aiming to foster discussion around racial justice and structural inequalities in the justice system (Reina 2021). However, as the letters were installed, the Kinderhook zoning board filed a stop order based on the artwork violating the city's regulations on public signage. The gallery had to pay a fine of 200 dollars for every day that *Truth to be told* remained in place. This stirred not only a legal battle but also a passionate debate online, in the village and beyond, about petty bureaucracy, censorship and undercover racism in institutions. The debate was strategic to foreground *Truth to be told* as an example of political art seeking to respond to current inequalities and misperceptions in the court hearing four months later.

In addition to these two examples of art creating publicness, the aesthetics and performative forms of coming together in activist experiments during pandemics, especially the ones taking place in Black and brown communities, are also inspiring for posthumanist arts-based interventions in the city. First, the masked protests after George Floyd's killing and the very act of walking together turned out to be moving assemblages that allowed the participants for political (semi)anonymity and gave rise to new affective attachments to the urban (Ticktin 2020). Additionally, mutual aid to counter the economic recession during lockdown is not only remarkable for reasons of solidarity or charity but as mentioned earlier, as a reworking of material infrastructures. This is the case of the *friendly fridges* set up across New York City (Rosa 2020). From February 2020 onwards, these fridges were sited at the sidewalks of several neighbourhoods. Commonly stocked with unused or unsold food items from restaurants, they welcomed anyone to take free food/*comida gratis*. No explicit agreement was needed between the network of volunteers in charge of the refrigerators and people who took what they wanted from them. The fridges not only provided free food without any ownership or leadership behind but also displayed a communal atmosphere in the city. It slightly reminds previous art projects such as Gaye Chan and Nandita Sharma's *Eating in public* (2003), where the participants were encouraged to plant and cultivate edible indigenous food on public lands like "the seventeenth-century English commoners (…) at the outset of the private-property revolution" (Ticktin 2020).

These practices evoke patterns of what care and relationality in urban space may entail in the state of current crisis and controlled circulation. They show that care is not just doing more as an individual to help those in need through charitable efforts. We see how governments and organizations continue subsuming such acts in the abstracting and appropriative logics of neoliberalism. What is needed is an "imagination of undisciplined politics of inhabitation" (Lancione and Simone 2020b, n.p.) that undoes the current fixed points in which we found ourselves. As Lancione and Simone (2020b) argue, if we aim for a collective life driven by care, then we cannot live in isolation and without risk. They write, "In order for care to take place, persons must extend themselves one another beyond the positions and sensibilities they occupy. Care must entail the capacity of persons to be implicated in a world beyond where it is" (n.p.). This is an opening for participatory arts-based research to extend and support creative, provocative, and undisciplined capacities of being in public in a time when being in public is under further control and threat but more in need than ever to be reassembled. This could be done through experiments dedicated to recalibrating our sensibilities, collectively reinhabiting sidewalks, buildings, aisles, playing off with ways to space out each other that generate intimacies, and propagate questionings and desires through newly re-materialized landscapes.

References

Amin, Ash and Nigel Thrift. 2017. *Seeing like a city*. Cambridge and Malden, MA: Polity Press.

Barad, Karen. 2003. "Posthumanist performativity: Toward an understanding of how matter comes to matter". *Signs: Journal of Women in Culture and Society* 28, no. 3: 801–831. https://doi.org/10.1086/345321.

Béhague, Dominique, William Minter and Francisco Ortega. 2021. "Solidarity, infrastructure and critical pedagogy during COVID-19: Lessons from Brazil". *Somatosphere*. http://somatosphere.net/2021/solidarity-infrastructure-and-critical-pedagogy.html/ (accessed June 10, 2021).

Behr, Rafael. 2021. "What does 'returning to normal' mean with a prime minister like Boris Johnson?" *The Guardian*, April 7, 2021. https://www.theguardian.com/commentisfree/2021/apr/07/returning-to-normal-prime-minister-boris-johnson-covid-politics

Berlant, Lauren. 2016. The commons: Infrastructures for troubling times. *Environment and Planning D: Society and Space* 34, no. 3: 393–419 https://doi.org/10.1177/0263775816645989

Bignall, Simone and Rosi Braidotti. 2019. "Posthuman systems". In *Posthuman ecologies: Complexity and process after Deleuze*, edited by Rosi Braidotti and Simone Bignall, 1–16. London: Rowman & Littlefield.

Butler, Judith. 2020. "The force of nonviolence." Whitechapel Gallery, 23 July, 2020. YouTube video, 1:07:02. https://www.youtube.com/watch?v=HN5D9rlkRcA.

Corsín Jiménez, Alberto. 2014. "The right to infrastructure: A prototype for open source urbanism". *Environment and Planning D: Society and Space* 32: 342–362. https://doi.org/10.1068/d13077p

Deleuze, Gilles and Felix Guattari. 1987. *A thousand plateaus. Capitalism and schizophrenia*. Minneapolis: University of Minnesota Press.

Ellsworth, Elizabeth. 2005. *Places of learning: Media, architecture, pedagogy*. New York: Routledge.

Ferreira da Silva, Denise. 2020. "The future of two presents". *Social Text*. https://social-textjournal.org/periscope_article/the-future-of-two-presents/ (accessed June 2, 2021).

de Freitas, Elizabeth, David Rousell, and Nils Jäger. 2020. "Relational architectures and wearable space: Smart schools and the politics of ubiquitous sensation". *Research in Education* 107, no. 1: 1–23. https://doi.org/10.1177/0034523719883667.

Haraway, Donna. 1988. "Situated knowledges: The science question in feminism and the privilege of partial perspective." JSTOR 14, no. 3: 575–599. https://www.jstor.org/stable/3178066.

Haraway, Donna. 2008. *When species meet*. Minneapolis, MN: The University of Minnesota Press.

Haraway, Donna. 2016. *Staying with the trouble. Making kin in the Chthulucene*. Durham: Duke University.

Haraway, Donna. 2020. "Phonocene." CCCB, 16 December 2020. YouTube video, 23:43. https://www.youtube.com/watch?v=87HzPIEiF78&t=1210s.

Kapadia, Jeevika and Ashwini Sirsikar. 2020. "Covid-19: the great leveller". *Ipsos*, April 24, 2020. https://www.ipsos.com/en-in/covid-19-great-leveller.

Lancione, Michelle and AbdouMaliq Simone. 2020a. "Bio-austerity and solidarity in the COVID-19 space of emergency – Episode one." *Society + Space*. https://www.societyandspace.org/articles/bio-austerity-and-solidarity-in-the-covid-19-space-of-emergency (accessed May 10, 2021).

Lancione, Michelle and AbdouMaliq Simone. 2020b. "Bio-austerity and solidarity in the COVID-19 space of emergency – Episode two." *Society + Space*. https://www.societyandspace.org/articles/bio-austerity-and-solidarity-in-the-covid-19-space-of-emergency-episode-2 (accessed May 10, 2021).

Larkin, Brian. 2013. "The politics and poetics of infrastructure". *Annual Review of Anthropology* 42: 327–343. https://doi.org/10.1146/annurev-anthro-092412-155522.

Latour, Bruno. 2014. "What is the style of matters of concern?" In *The lure of whitehead*, edited by Nicholas Gaskill and A.J. Nocek, 92–126. Minneapolis: University of Minnesota Press.

Massumi, Brian. 2002. *Parables of the virtual: Movement, affect, sensation*. Durham, NC: Duke University Press.

McCormack, Derek P. 2018. *Atmospheric things: On the allure of elemental envelopment*. Durham: Duke University Press.

OECD. 2021. "OECD Green recovery database". https://www.oecd.org/coronavirus/policy-responses/the-oecd-green-recovery-database-47ae0f0d/ (accessed June 9, 2021).

Price, Catherine. 2020. "Covid-19: When species and data meet". *Postdigital Science and Education* 2: 772–790. https://doi.org/10.1007/s42438-020-00180-x

Reina, Barbara. 2021. "In unanimous decision, Kinderhook zoning board recognises Nick Cave's Truth Be Told is art not a sign". *The Art Newspaper*, February 5, 2021. https://www.theartnewspaper.com/news/in-unanimous-decision-kinderhook-zoning-board-recognises-nick-cave-s-truth-be-told-is-art-not-a-sign (accessed June 10, 2021)

Rosa, Amanda. 2020, July 8. "See that fridge on the sidewalk? It's full of free food". *New York Times*. https://www.nytimes.com/2020/07/08/nyregion/free-food-fridge-nyc.html (accessed June 21, 2021).

Roy, Arundhati. 2020. "The pandemic is a portal". *The Financial Times*, April 3, 2020. https://www.ft.com/content/10d8f5e8-74eb-11ea-95fe-fcd274e920ca (accessed April 15, 2020).

Satvrides, Stavros. 2020. "Life in commons". *Undisciplined Environments*. https://undisciplinedenvironments.org/2020/05/08/life-as-commons/ (accessed June 9, 2021).

Shaviro, Steven. 2012. *Without criteria: Kant, whitehead, deleuze, and aesthetics*. Cambridge, MA: MIT Press.

Simone, AbdouMaliq. 2020. "The pandemic, southern urbanisms and collective life". *Society + Space*. https://www.societyandspace.org/articles/the-pandemic-southern-urbanisms-and-collective-life (accessed May 8, 2021).

Solnit, Rebecca. 2020. "The way we get through this is together': The rise of mutual aid under coronavirus". *The Guardian*, May 14, 2020. https://www.theguardian.com/world/2020/may/14/mutual-aid-coronavirus-pandemic-rebecca-solnit (accessed May 25, 2020).

Ticktin, Miriam. 2020. "Building a Feminist Commons in the Time of COVID-19". *Signs Blog*. http://signsjournal.org/covid/ticktin/ (accesed June 21, 2021).

Whitehead, Alfred North. 1964. The concept of nature. Cambridge: Cambridge University Press.

World Health Organization. 2021. "WHO-convened global study of origins of SARS-CoV-2: China part". Joint WHO-China study. https://www.who.int/publications/i/item/who-convened-global-study-of-origins-of-sars-cov-2-china-part (accessed June 10, 2021).

Index

Page numbers in *italics* refer to figures and page numbers followed by 'n' refer to notes numbers.

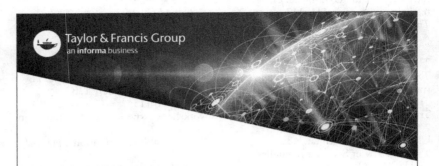

Printed in the United States
by Baker & Taylor Publisher Services

Printed in the United States
by Baker & Taylor Publisher Services